Emergency Care for Birds

Due to their often small size and unique physical characteristics, birds can deteriorate rapidly in the event of illness and injury. Timely intervention in the event of clinical signs is therefore essential for an optimal prognosis. Authored by avian veterinarian Rob van Zon, this concise, practical guide will help veterinary professionals to provide first aid and emergency care for birds, as well as to instruct clients on providing basic first aid when they are unable to get to the clinic.

This full-color book, packed with photographs and anatomy drawings, includes instructions for veterinary professionals on stabilizing sick avian patients and management of many specific emergency situations at the veterinary clinic. This includes guidance for those presented with a moribund, critically ill patient i.e., recognizing and treating hypothermia, hypoxia, hypovolemia and hypoglycemia. The book also lists alarming signs of disease and toxic plants, and includes clinical procedures, an emergency drug formulary, and bandaging techniques. Finally, it offers guidance to veterinarians giving advice to bird owners for first aid at home.

Emergency Care for Birds

A Guide for Veterinary Professionals

Rob van Zon

with contributions and edits from

Dr. Ineke Westerhof (ECZM-avian)

CRC Press
Taylor & Francis Group
Boca Raton New York London

CRC Press is an imprint of the
Taylor & Francis Group, an **informa** business

First edition published 2024
by CRC Press
6000 Broken Sound Parkway NW, Suite 300, Boca Raton, FL 33487-2742

and by CRC Press
4 Park Square, Milton Park, Abingdon, Oxon, OX14 4RN

CRC Press is an imprint of Taylor & Francis Group, LLC

ISBN: 978-1-032-31141-8 (hbk)
ISBN: 978-1-032-31132-6 (pbk)
ISBN: 978-1-003-30827-0 (ebk)

DOI: 10.1201/9781003308270

Typeset in Palatino
by KnowledgeWorks Global Ltd.

Contents

Acknowledgments

Special thanks to Dr. Ineke Westerhof (ECZM-avian), Zoe van der Plaats, Ellen Rasidi, Dr. Ann Bourke, Crissy Olson, Mandy Wong, Alvaro Guzman and Bianka Schink.

About the author

Rob van Zon

Having been interested in birds throughout his youth, Rob van Zon started studying veterinary medicine at the University of Utrecht to turn his passion into his job. After graduating in 2005, Rob worked as an avian veterinarian in veterinary practices in Amsterdam and Utrecht and in avian wildlife centers in the Netherlands. In addition to treating thousands of birds in his own clinics, Rob tries to help as many birds as possible indirectly by teaching other veterinarians, students and bird owners.

Introduction

This book was written to help general practice veterinarians dealing with medical emergencies affecting birds.

Birds are popular pets, and rightly so. Because of their high intelligence, unique character and beautiful appearance, birds truly deserve their place in and around our homes. Unfortunately, living in a man-made environment does carry risks for our avian companions. The evolutionary adaptations that are useful for survival in nature do not protect our birds from the dangers present in the average household. For example, birds would not encounter electrical wires, overheated non-stick pans, bleach or ceiling fans in their natural environment. Furthermore, birds in nature can learn what is safe to eat from their peers, warn each other about danger and flee from other birds in case of an argument.

Due to their often small size and unique physical characteristics, birds can deteriorate rapidly in the event of illness and injury. Timely intervention in the event of clinical signs is therefore essential for an optimal prognosis.

Most general-practice veterinarians work primarily with dogs and cats and may have little experience in avian medicine. Scientific and medical fields have developed and expanded so rapidly over the last decades that even in human medicine (treating just one species), clinical specialists only practice in their specialist field. The same now applies to veterinary medicine, where it is impossible for veterinarians to have detailed knowledge of all animal species. Therefore, avian medicine and surgery have developed into a separate clinical specialty, and many experienced avian specialists are available to provide veterinary care. Unfortunately, for some bird owners, these avian specialists might be too far away. Moreover, while consultation with an avian specialist may be possible during office hours, this might not be the case at night, on weekends or on holidays. Because of those circumstances, owners of sick or injured birds might present to a general practice veterinarian for initial treatment and care, and these veterinarians may be less experienced and knowledgeable in avian medicine.

Part 1 of this book describes acute clinical signs of disease and why birds showing these symptoms should be seen by a veterinarian as soon as possible.

Part 2 contains information about handling emergencies in birds and how to stabilize the sick avian patient.

Specific emergency situations are discussed in Part 3, which contains both first aid-advice for owners at home and suggestions for veterinarians during emergency consultations.

This book also gives suggestions for examination and treatment in certain situations. Alternative approaches may be taken and are not necessarily wrong. Vets with experience in avian medicine, or other diagnostic options, often have their own methods that can also result in positive outcomes.

Some of the treatment recommendations are not authorized for use in all geographical regions and some may not be authorized for use in all bird species. It is the treating veterinarian's duty to make a risk assessment for each patient prior to administering any treatment.

Part 1: Acute clinical signs of disease

1 *Clinical signs of disease*

UNLIKE dogs and cats, birds are animals that are not evolutionarily adapted to cohabitation with humans. Birds are not domesticated. In most cases, our pet birds are genetically the same as their wild family members. As a result, they also show the same innate behavior and traits, one of which is that they instinctively try to hide signs of disease. In nature, this prevents attracting the attention of predators. Unfortunately, in captivity, this same instinctive behavior of hiding symptoms can lead to the situation where physical problems are only noticed by bird owners in an advanced stage of disease. Even with serious illnesses, symptoms can initially be masked and unnoticed.

Signs of illness in birds include abnormalities in breathing, skin, behavior, appetite, droppings, feathers, posture, coordination, swellings, nausea, eyes, ears, nose, vent and body weight. Any changes may be considered relevant. As soon as signs of disease are evident, there is every reason to consult an avian veterinarian. In some cases, immediate examination and treatment are necessary, regardless of the time of day (or night). In other cases, the visit to the veterinarian could be postponed for up to 24 hours or longer. If birds show any of the symptoms listed in Chapter 2, seeking medical care as soon as possible is recommended.

DOI: 10.1201/9781003308270-2

2 *Signs of disease requiring immediate attention by a veterinary professional*

- Repeated vomiting or regurgitation
- Sudden decrease of appetite or anorexia
- Change in behavior
- Sitting still with fluffed feathers
- Changed voice
- Faster and/or heavier breathing, abnormal breathing sounds (dyspnea)
- Sudden swelling of any part of the body
- Deep injury or blood loss
- Acute change in the amount or appearance of the droppings (unless explained by a change of diet)
- Discharge from the beak (from the "mouth", not from the nose)
- Straining to lay, or visible egg in the cloaca
- Altered position of a wing, leg or head
- Sudden lameness
- Contact with or ingestion of a toxic substance
- Seizures/fits
- Fainting
- Ataxia/loss of coordination
- Paralysis
- Cloacal prolapse
- Abnormal eye or inability/unwillingness to open the eye(s)
- Death of other birds living in close contact
- Too much watery urine (polyuria)
- Increased water intake (polydipsia)
- Change in body weight or body condition
- Nasal discharge
- Eye discharge

BACKGROUND AND EXPLANATION

REPEATED VOMITING OR REGURGITATION

Vomiting in birds is often a more violent action in which the contents of the stomach/crop are expelled out the mouth. The bird may often appear to flick fluid around the immediate environment, and the head feathers often appear wet or stained with vomitus. Regurgitation, on the other hand, is a less violent action

DOI: 10.1201/9781003308270-3

in which food from the throat or crop appears to "spill over" from the beak and generally results in less mess.

Vomiting can be a symptom of a relatively harmless condition, but also of acute life-threatening illnesses, such as intoxication, ingested foreign bodies, gastrointestinal obstruction and liver or kidney failure. Besides the fact that vomiting can have a serious cause, it can also have serious consequences. The loss of too much food or water can quickly lead to complications from starvation or severe dehydration.

Note: When a bird actively gives up ingested food (regurgitation) aimed at the partner, owner, toy, mirror or other object, this is probably "partner behavior". In this case, there is no vomiting or threatening emergency situation. However, repeated regurgitation in the absence of a "recipient" may indicate underlying infection or other issue.

SUDDEN DECREASE OF APPETITE OR ANOREXIA

Birds have a high body temperature and a very fast metabolism. A sudden or acute decrease in food intake leads to an energy shortage and serious metabolic problems. In addition, stopping eating can lead to life-threatening bleeding in the intestines (in small birds within 24 hours). Regardless of the cause, eating too little or stopping eating completely can quickly lead to a life-threatening situation.

CHANGE IN BEHAVIOR

Behavioral changes can indicate non-specific signs of disease. With many illnesses, birds become less active or even lethargic. In other cases, for example certain intoxications, they can also become hyperactive instead or act very irritated. In case of sudden obvious changes in behavior, it is therefore advisable to have the bird examined immediately.

SITTING STILL WITH FLUFFED FEATHERS

Sitting still with fluffed feathers is a non-specific symptom of illness, pain or cold. It can be caused by very serious diseases, so it is advisable to have the bird examined immediately.

CHANGED VOICE

Voice can be changed due to very serious problems in the trachea and/or the vocal organ (syrinx). A notorious cause is the fungal infection aspergillosis. Since a total blockage of the airways with asphyxiation can sometimes be the next stage of the disease, it is important to have a bird with an abnormal or lost voice examined immediately.

FASTER AND/OR HEAVIER BREATHING, ABNORMAL BREATHING SOUNDS

Faster and/or forced or heavy breathing, or abnormal breathing sounds are symptoms of dyspnea, or shortness of breath. Shortness of breath can be caused by diseases in the airways themselves (for example lung or air sac diseases), but may be also due to cardiovascular diseases, neurological disorders, fluid in the coelom/abdomen or a space-occupying mass in the coelom/abdomen, such as an egg, enlarged organ or tumor. Difficulty breathing may result in oxygen deficiency, which can quickly lead to dangerous situations, so it is wise to seek veterinary attention immediately.

SUDDEN SWELLING OF ANY PART OF THE BODY

Rapidly occurring swelling of the limbs or head often has a serious cause, for example severe inflammation, leg band constriction, fracture, dislocation or hematoma. Rapid action is often required to prevent serious complications. For example, in the case of leg band constriction, the leg band should be removed as soon as possible to prevent necrosis (death) of the foot. Swelling of the coelom/abdomen is often accompanied by shortness of breath or dyspnea.

DEEP INJURY OR BLOOD LOSS (EXCEPT FOR A SINGLE DROP)

Excessive blood loss quickly leads to problems with blood pressure, resulting in shock or even death. For a small bird the size of a budgerigar, the loss of more than five drops of blood is already threatening.

ACUTE CHANGE IN THE AMOUNT OR APPEARANCE OF THE DROPPINGS (UNLESS EXPLAINED BY A CHANGE IN DIET)

A significant reduction in the amount of feces may indicate inadequate food intake, which can lead to serious metabolic problems and the occurrence of spontaneous bleeding from the gut wall (usually life-threatening).

Red or black discoloration of the feces may indicate bleeding from the gastrointestinal tract. When bleeding occurs in the last part of the intestines, rectum or cloaca, bright red blood is usually visible in the stools. Bleeding from the higher part of the gastrointestinal tract usually results in black, tarry feces. Very pale feces may indicate a digestive problem and be a symptom of severe pancreatic disease, which may be caused by infections or intoxications.

Diarrhea can have multiple causes and can lead to dehydration and loss of electrolytes and nutrients.

A change in color of the normally white uric acid crystals (urates), or clear urine, is usually the result of serious illness. Yellow or green discoloration can indicate damage to the liver. Pink discoloration can occur with very serious kidney damage and certain intoxications.

Note: If the stool is no longer fresh, urates can discolor to slightly green due to mixing of the dyes from the feces.

DISCHARGE FROM THE BEAK

Discharge from the beak (from the "mouth", not from the nose) can have several causes, including life-threatening illnesses or extreme weakening.

STRAINING OR VISIBLE EGG IN CLOACA

Straining can be a sign of egg binding (dystocia). Egg binding is the situation in which a bird fails to lay a fully or partially formed egg. In some cases of egg binding, the egg is actually visible in the cloaca, but in most cases this isn't so. Egg binding can lead to dangerous complications, such as damage to the kidneys and nerves, cloacal prolapse, constipation, shortness of breath, and reduced food and water intake. If egg binding is not treated adequately in time, it can lead to the death of the bird.

ALTERED POSITION OF A WING, LEG OR HEAD

This often indicates an orthopedic problem (such as a fracture or dislocation), pain, tumor or a neurological problem. Late medical intervention can lead to dangerous complications or permanent injuries.

SUDDEN LAMENESS

Sudden lameness is usually the result of a serious orthopedic abnormality (such as a fracture or dislocation), leg band constriction, intoxication, pain, tumor or neurological problem.

CONTACT WITH OR INGESTION OF A TOXIC SUBSTANCE

When it is known that a bird had been in contact with a poisonous substance—for example lead, zinc, poisonous plants, chocolate, medications, avocado, and mouse or rat baits (rodenticides)—action must be taken as quickly as possible to attempt to prevent problems or even death. This can be done, for example, by lavaging the crop, laxative treatment, protection of mucosal linings, fluid therapy or preventive treatment with an antidote. The sooner this is done, the better the prognosis.

SEIZURES/FITS

A seizure is characterized by cramps, shocks and uncontrolled spasms, often accompanied by reduced consciousness. Epilepsy is a neurological condition causing repeated seizures. Although epilepsy does not always have a threatening cause, a seizure can be a sign of a life-threatening condition. For example, intoxication, cardiovascular disease, brain disease, liver or kidney failure, and calcium or glucose deficiency can cause seizures. Often the underlying cause of the seizure is also causing damage to other parts of the body. In addition, prolonged or repeated seizures can lead to overheating and death of the patient. If a bird has not already previously been diagnosed with and not already being treated for epilepsy, a seizure is always a reason to seek medical help as soon as possible.

FAINTING

Fainting usually has a serious cause, e.g., heart failure. It is therefore always a reason to consult a vet as soon as possible.

ATAXIA/LOSS OF COORDINATION

A loss of coordination ("drunken gait", slips, misses or falling off the perch) is due to reduced function of the nervous system. Notorious causes are intoxications (for example lead poisoning), brain trauma, infections of the brain, cerebral infarction (stroke) or brain tumors.

PARALYSIS

Paresis (weakness) or paralysis (loss of function) of the legs can occur with egg binding or other reproductive disease, trauma, intoxication, tumor or kidney disease (swollen kidneys can press on the nerve that runs from the spinal cord to the hind leg). Botulism is a notorious cause of paralysis in waterfowl. Rapid medical intervention will lead to a better chance of recovery and reduce the chance of permanent damage in many cases.

CLOACAL PROLAPSE

In the case of a prolapse, the inside of the cloaca, and/or part of the distal intestine and/or reproductive tract hangs out at the vent. Regardless of the cause, a prolapse quickly leads to potentially life-threatening complications. The mucous membranes can dehydrate, swell, become damaged or become infected, and there may be unseen internal injury as well. This situation can quickly lead to a life-threatening situation or cause irreversible damage.

ABNORMAL EYE OR INABILITY/UNWILLINGNESS TO OPEN THE EYE(S)

The eyes may be abnormal or closed in the event of very severe general malaise or weakening, or in abnormalities of the eyes or eyelids. These include infections, tumors, damage to the eyeball (globe), presence of a foreign object under the eyelid, allergic reactions or insect stings. In the latter cases, complications can quickly develop. The most severe cases can lead to permanent damage to the eye and blindness.

DEATH OF OTHER BIRDS LIVING IN CLOSE CONTACT

Death in other birds may indicate a severe infectious disease or exposure to toxin.

TOO MUCH WATERY URINE (POLYURIA)

In birds, droppings consist of feces, urates (uric acid crystals) and watery urine. During stress, after bathing and after eating water-rich foods such as fruit, it is normal for more urine to be produced. Pathological causes of polyuria include kidney or liver disease, intoxication, diabetes or other endocrine/hormone diseases and psychogenic polydipsia. When an increased loss of fluid through the urine is not compensated by increased water intake (polydipsia), it can lead to dehydration.

INCREASED WATER INTAKE (POLYDIPSIA)

Causes of an increased water intake (drinking more) include kidney or liver disease, intoxication, diabetes or other endocrine/hormone diseases and psychogenic polydipsia. Increased water intake is often accompanied by an increase in clear, watery urine (polyuria) in the droppings.

CHANGE IN BODY WEIGHT OR BODY CONDITION

A healthy bird maintains a fairly constant weight and body condition. Body condition in birds is assessed by the mass of the pectoral muscles (the "breast fillets" on a chicken). Weight loss may be due to a decreased appetite, poor digestion, loss of nutrients, increased energy requirements or dehydration. Weight loss is a non-specific sign of many diseases.

Note: In some pathological conditions, body weight may remain the same or even increase. Examples are diseases in which free fluid accumulates in a body cavity or in which cysts or tumors grow in the body. In these cases, body condition declines despite the body weight remaining the same or potentially increasing, resulting in a loss of muscle mass. This is best noted by checking the mass or volume of the pectoral muscles. A "sharp" chest indicates reduced body condition and is often a sign of illness.

NASAL DISCHARGE

Nasal discharge is fluid coming out of the nostrils. This fluid can be watery, bloody, slimy, or pus-like. Sometimes, it can be subtle, so that the only sign is a discoloration of the feathers around the nostrils. Nasal discharge can occur with an abnormality in the nose or sinuses.

EYE DISCHARGE

Eye discharge from a normally open eye can occur, for example, with a blocked tear duct or conjunctivitis.

Part 2: Birds as emergency patients at the veterinary clinic

3 Observation, physical examination and diagnostic tests

A THOROUGH external physical examination is just as important when treating birds as it is when treating mammals.

OBSERVATION FROM A DISTANCE

Before handling the patient, it is important to observe the bird first from a distance and examine the cage.

In the cage, useful information can be seen, such as the appearance of the droppings, evidence of vomiting (**Fig. 3.1**), type of food offered or potential hazardous materials.

Some time should be taken to observe the patient from a distance (staring directly at the bird like a predator should be avoided). It must be kept in mind that most birds will try to mask signs of disease and will not show abnormalities immediately.

The bird is evaluated for:

- *Mental status*: Depression, sitting fluffed with closed eyes, sitting on the floor instead of on a perch.
- *Posture, position*: of head (head tilt, opisthotonus), position of the limbs, ataxia, paresis, paralysis, seizures, spams, tremors, involuntary movements or sounds ('tics').
- *Respiratory rate:* breathing sounds and signs of labored breathing (tail bobbing, opened-mouthed breathing).

DOI: 10.1201/9781003308270-5

Fig. 3.1 Soiled feathers on top of the head of a budgerigar and seeds sticking to the cage indicating nausea.

Note: Birds that are severely dyspneic, debilitated or in shock should be stabilized before undergoing a stressful examination. *See Chapter 10, p. 39.*

HANDLING

To examine and treat a bird, it is necessary to handle the patient. When handling an avian patient, it is important to immobilize the bird (to avoid damage to both the patient and staff), but minimize stress and avoid restricting respiration by putting as little pressure on the bird's body as possible. Windows and doors should be closed before removing the bird from the cage/transport box. It is also important that windows are covered or blinded to avoid a hard collision by an escaped patient.

Handling is not entirely without risk in seriously ill, weakened or dyspneic birds. See Appendix 1, p. 161, for the technique of handling different bird species.

PHYSICAL EXAMINATION

BODY WEIGHT

Tame birds can be weighed directly on a scale (**Fig. 3.2**) or on a perch on a scale (**Fig. 3.3**). Birds that tend to fly away can best be weighed in a container on a scale.

Fig. 3.2 African grey parrot on scale.

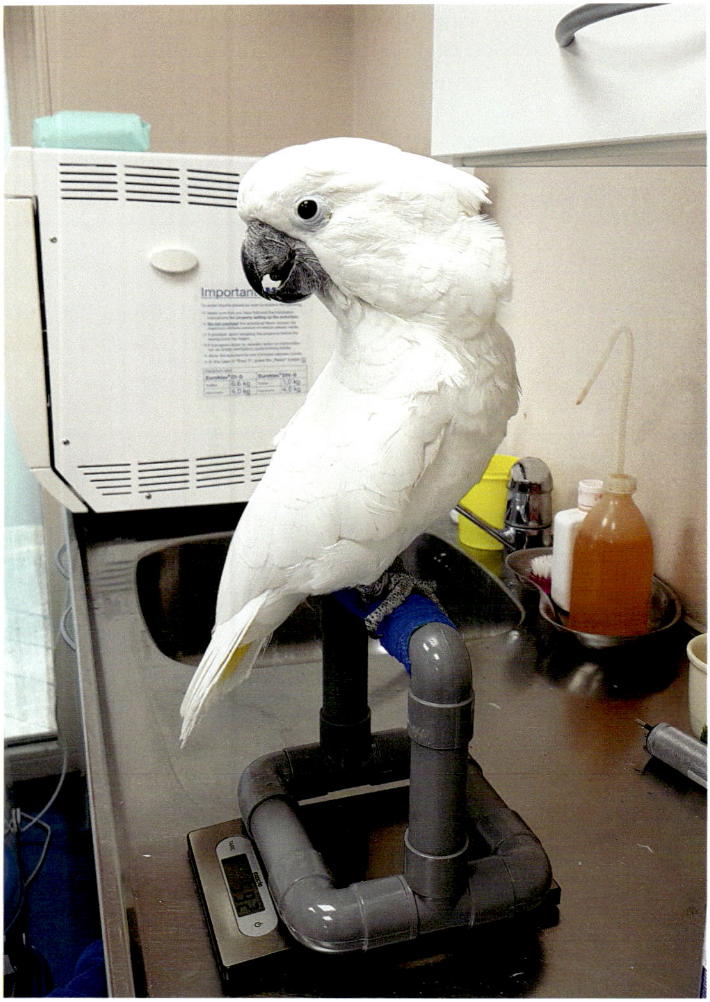

Fig. 3.3 Sitting on a perch on a scale can be better accepted (and more practical in cases of tame birds with long tails).

EYES

Check for discharge, injuries, conjunctiva swelling, foreign objects, nystagmus (the normal eye pinning or eye flashing caused by excitement should not be mistaken for involuntary nystagmus), cornea transparency, color and transparency of the aqueous humor in the anterior chamber, color and symmetry of the iris and opacity of the lens. Check skin turgor in the upper eyelid.

NARES

Check for exudate, asymmetry and obstruction. The operculum (a normal structure in the nares of many species) should not be confused for a foreign object or abnormal structure. When in doubt about normal anatomy of a nostril, the other nostril can be used for comparison.

BEAK

Check for injuries, overgrowth, deformities, asymmetry, abnormal patterns of wear, dark spots and softness of keratin.

ORAL CAVITY

Check for discoloration, lesions, swelling and ulceration or plaques of mucous membranes, exudate/mucus and foreign objects.

EARS

Check for exudate, redness and swelling of the skin.

CROP

Check for filling/distension, consistency of contents, foreign bodies, swelling of the wall and discoloration of the skin over the crop.

CONTOUR FEATHERS

Check for discoloration, damage, loss of feathers, abnormal molting, parasites and stress lines/fret marks.

SKIN

Check for injuries, ulceration, ectoparasites, masses, redness and flaking.

KEEL BONE, DEVELOPMENT OF PECTORAL MUSCLES AND BODY CONDITION SCORE (BCS)

Check the development of pectoral muscles and the amount of subcutaneous fat to determine the Body Condition Score (BCS).

Muscle development, and thereby the prominence of the keel bone (Carina), gives an indication of the nutritional status of the avian patient.

Note: Normal anatomy and muscle development vary between species and breeds. For example, flightless birds have less well-developed pectoral muscles than flying birds. In chickens, laying breeds have less well-developed pectoral muscles than meat breeds.

For most flying birds, the muscles should be just rounded and the keel should be easy to palpate and visible after wetting the skin. The muscles should not protrude the edge of the keel bone and no layer of fat (soft, yellow tissue) should be present between the keel bone/muscles and the skin.

BCS is based on the muscular development and on the amount of subcutaneous fat in the neck, chest and pelvic area.

The BCS can be very helpful in emergency situations to differentiate between acute and more chronic diseases. Birds with low BCS are more likely to suffer from chronic diseases than birds with high BCS. Of course, comorbidity can exist, so a low BCS doesn't rule out an acute problem.

AUSCULTATION

The heart is auscultated at the thoracic inlet and at both sides of the ventral thorax over the pectoral muscles. Check for murmurs, heart rate and arrhythmias. The lower airways are auscultated at the lateral and dorsal body wall. Check for abnormal breathing sounds.

Note: Normal auscultation does not rule out respiratory or cardiovascular disease.

ABDOMINAL PALPATION

The coelom/abdomen is palpated between the caudal edge of the sternum and the cloaca. Check for distension and masses. In healthy birds, the only firm organ that can be palpated is the ventriculus. The presence of eggs can be a sign of normal reproduction or egg binding.

CLOACA

Check for injuries, prolapse, paralysis, swelling, masses, eggs, cloacoliths and soiling of surrounding feathers. In poultry, gentle cloacal palpation can be used for examining the caudal coelom.

LEGS

Check for asymmetry, swelling and masses. Check the skin of the legs and especially the plantar side of both feet for injuries, erosions, hyperemia and ulceration. Check all the joints of the legs and feet for range of motion, swelling, crepitation, painfulness. Check all the bones for crepitation, abnormal mobility, pain and deformities. Check the nails for injuries and length. Check the possibly present legband for the age of the bird and ensure it is not too tight. Check skin turgor on the dorsal side of the feet.

TAIL

Check for discoloration, damage, loss of feathers, abnormal molting, parasites and stress lines/fret marks.

UROPYGIAL GLAND

Check the papilla and overlying skin for swelling, redness and ulceration. Check the lobes of the uropygial gland for asymmetry, distension and pain. When distended, gentle pressure can be applied on the lobes trying to express some of the oily excretion to check for blockage of the ducts. Some birds do not have a uropygial gland, for example Amazon parrots and some macaws, doves and pigeons.

WINGS

Check for asymmetry, swelling and masses. Check the skin for injuries and ulceration. Check all the joints for range of motion, swelling, crepitation and painfulness. Check all the bones for crepitation, abnormal mobility, pain and deformities. Check the feathers of the wings for discoloration, damage, loss of feathers, abnormal molting, parasites and stress lines/fret marks. Check the filling and refill time of the basilic vein (see Fluid therapy, p. 27).

BACK

Check the skin for injuries, swelling and masses. Check the spine for deformities. Check the contour feathers for discoloration, damage, loss of feathers, abnormal molting, parasites and stress lines/fret marks.

After handling posture, mental status and respiration are re-evaluated. Healthy birds fully recover from stress from handling within 2 minutes.

Unfortunately, it is often not possible to diagnose or exclude internal abnormalities just by inspection, palpation and auscultation. More diagnostic tests are frequently necessary.

EXAMINATION OF DROPPINGS

Avian droppings normally consist of feces, uric acid crystals (urates) and watery urine. The first step is macroscopic examination of the droppings: The amount, color and consistency of the feces, presence of worms or blood, the amount of watery urine and the color of the urates are determined. The second step is microscopic examination of fecal samples. Wet mounts and stained smears are examined for the presence of blood, inflammatory cells, parasites, yeasts and bacteria. See Appendix 10, p. 217.

RADIOLOGY

Radiology is an essential part of diagnostics in birds.

X-rays, ultrasound and computed tomography (CT)-imaging are useful in avian practice, all with their own advantages and disadvantages.

ULTRASOUND

Ultrasound can be used in birds to examine the liver, heart, kidneys, reproductive system and soft tissue swellings in the coelom and can also be used to diagnose ascites or egg binding.

A disadvantage of ultrasound in birds is that the small body size, high respiratory rate and the presence of air sacs can limit the image quality and diagnostic value. Another disadvantage is that the pressure applied with the probe can aggravate pre-existing respiratory distress.

X-RAYS

X-rays are the most widely used method of radiology in avian practice. X-rays are used for visualizing bones, air sacs, internal organs and other soft tissues and for determining the presence of ascites and radiodense structures (e.g. metal particles, calcified egg shells, cloacoliths and calcified tissues). Disadvantages of X-rays include the superimposition of tissues and the necessity of anesthesia/sedation for optimal patient positioning and safety for the patient and staff when taking X-rays of birds.

See Appendix 3, p. 179.

COMPUTED TOMOGRAPHY

CT takes away the disadvantages of tissue superimposition and is superior to X-rays in visualizing bones, sinuses and the lower respiratory tract. CT is also used for visualizing the soft tissues of the body. Disadvantages of CT include costs, the necessity of anesthesia and the relatively low resolution of soft tissues, especially in small species.

ENDOSCOPY

Endoscopy with a rigid endoscope is a practical technique in avian practice for visualizing the trachea and syrinx, air sacs and abdominal organs (the site and technique for entering the air sacs is the same as for the placement of an air sac tube, see Appendix 6, p. 201), cloaca and the esophagus, crop and stomach. Endoscopy can be helpful in the removal of foreign objects from the proximal gastrointestinal tract or trachea.

BLOOD TESTS

In healthy birds, up to 1% of body weight can be collected. In birds with dehydration/hypovolemia, up to 0.5% of body weight can safely be collected in most cases. See Appendix 2, p. 169 for accessible veins and technique.

Note: In very small species or seriously ill birds, collecting blood is sometimes too dangerous, especially by veterinarians less trained in collecting blood from birds. Too much blood loss, formation of hematoma, handling for too long or excessive stress can lead to death in small or unstable patients. In some cases, blood collection is best not done during the emergency consultation.

HEMATOLOGY

Hematology includes a complete blood count (CBC) and microscopic examination of a stained smear. These tests are helpful for diagnosing anemia, polycythemia and infection and for detection of blood parasites.

BLOOD CHEMISTRY

Blood chemistry tests are used to measure the amounts of certain chemicals in the blood, including enzymes, electrolytes, glucose, proteins, fats and minerals. A comprehensive blood chemistry panel in birds includes AST, GLDH, CK, BA, UA, BUN, TP, Alb, Glob, Ca, Glu and K, all described in the following sections.

Aspartate aminotransferase (AST)

AST is not an organ-specific enzyme. Increased AST can be the result of, for example, damaged cells of the liver or muscles.

Glutamate dehydrogenase (GLDH)

Increased GLDH is the result of severe damage/necrosis of cells of the liver. GLDH is the most specific enzyme for the detection of damage to liver cells, meaning that increased GLDH levels are very indicative of liver disease. Unfortunately, GLDH testing is not very sensitive, meaning that normal GLDH levels do not rule out damage to the liver cells.

Creatine kinase (CK)

Increased CK indicates damage to muscle cells. This can, for example, be caused by seizures, handling, trauma, stress or intramuscular injections.

Bile acids (BA)

Increased BA indicate a decreased liver function. Normal BA do not rule out liver disease. In birds lacking a gall bladder, for example pigeons and psittacines (except cockatoos), BA are most reliably measured after a 12-hour fasting period.

Note: Fasting is not advised in small species or birds that are already in a catabolic state.

Differentiating between liver damage and muscle damage is not possible based only on AST. For this reason, CK is measured as well. Increased AST without increased CK is indicative for liver cell damage. (Because of the shorter half-life of CK, older muscle cell damage can also cause this combination.) Increased AST combined with increased CK is indicative for muscle cell damage, although concurrent liver cell damage is not ruled out.

GLDH ↑ and / or BA ↑: Liver disease
AST ↑, CK ≈: Liver disease is likely
AST ↑, CK ↑: Muscle cell damage; liver disease is not ruled out

Uric acid (UA)

Increased UA in non-carnivorous species or fasted (24 hours) carnivorous species indicates renal failure.

Blood urea nitrogen (BUN)

Increased UN indicates severe dehydration. UN is not a sensitive marker of kidney failure in birds, meaning that normal or low UN does not rule out kidney failure.

For differentiating between acute kidney failure caused by dehydration and other causes of kidney failure, BUN, packed cell volume (PCV) TP and Alb can be measured besides UA. Increased BUN, PCV, TP and Alb are indicative of dehydration. Oliguria (low urine output) is common in this situation. Decreased or normal PCVand TP are indicative for other causes of kidney failure. Dehydration and other causes of kidney failure can exist independently or concurrently.

Total protein (TP)

TP consists of albumin and globulins. For interpretation of abnormal TP levels, differentiating between these two types of blood proteins is essential.

Albumin

Decreased albumin can be caused by liver disease, kidney disease, intestinal disease, (chronic) infections and malnutrition/starvation.

Increased albumin indicates dehydration and/or reproductive activity in females.

Globulins

Increased globulins indicate inflammation/infection.

Calcium

Total blood calcium consists of calcium bound to proteins, calcium bound to minerals and ionized calcium.

Note: Abnormal total calcium is in most cases not caused by problems with the calcium homeostasis. Decreased total calcium can, for example, be caused by hypoalbuminemia and increased total calcium by dehydration or female reproductive activity or dehydration.

Only the free and active ionized calcium fraction is involved in calcium homeostasis. Decreased ionized calcium indicates clinically relevant hypocalcemia. Increased ionized calcium indicates clinically relevant hypercalcemia, caused by, for example, vitamin D intoxication, excessive supplementation of calcium and neoplasia.

Glucose

Increased glucose can be caused by stress or diabetes mellitus. Decreased glucose can be caused by, for example, anorexia/starvation, chronic disease and sepsis.

Potassium (K)

Increased K can be caused by kidney disease and acidosis. Decreased K can be caused by excessive loss of potassium (diarrhea, vomiting and polyuria), insufficient uptake and alkalosis.

For clinical reference intervals of selected species, see Appendix 16, p. 235.

4 *General stabilization of sick birds*

BIRDS are animals with a fast metabolism, high body temperature and a relatively big surface area in comparison to their body weight, which favors heat loss. In addition, most birds are prey animals and hide symptoms of disease for as long as possible. As a result, birds presenting to the veterinarian are often already compromised. Most clearly sick birds are at least 5–10% dehydrated. In addition to dehydration (which can lead to hypovolemia and acute renal failure), hypothermia, hypoglycemia and hypoxia are often life-threatening.

Sick birds often have an urgent need for heat, fluids, nutrition and oxygen because of these factors. In many cases, sick birds presented as emergency patients can be stabilized by meeting these needs.

This indicates that in events of illness of unknown cause, sufficient time can often be gained by providing heat, fluids, nutrition and oxygen, so that the patient can be safely examined more thoroughly or be referred to an avian veterinarian later on.

DOI: 10.1201/9781003308270-6

5 *Heat*

THE AMBIENT temperature must be raised to an appropriate level by means of an external heat source such as a heated incubator, heat lamp or heat mat. The optimal temperature varies by bird species, age and case presentation.

AMBIENT TEMPERATURE GUIDELINE FOR SICK BIRDS

- *Adult birds (except waterfowl, meat-type chickens and birds from colder climates)*: 25–28°C
- *Young birds and bald adults*: 30–32°C
- *Birds with brain trauma, waterfowl, meat-type fowl and birds from colder climates*: room temperature

Note: Especially with psittacines, ensure that birds cannot bite on power cords or heat mats. Always make sure that a bird cannot reach or be burned by the external heat source.

DOI: 10.1201/9781003308270-7

6 *Fluid therapy*

FLUID therapy is indicated for resuscitation, rehydration and maintenance. Resuscitation is done in birds in (hypovolemic) shock and rehydration is relevant for dehydrated birds and maintenance for every patient.

FLUIDS USED FOR FLUID THERAPY

Crystalloids are composed of water and electrolytes. When used as plasma volume expanders by intravenous or intraosseous infusion, bigger volumes are needed in comparison to colloids, as about 70% of the administered volume will leave the intravascular space and move into the tissues. Crystalloids can be used for oral, subcutaneous, intravenous and intraosseous fluid therapy.

Colloids are composed of water, electrolytes and big oncotically active molecules. The big molecules stay in the intravascular space, thereby preventing water from moving into the tissues. Colloids are used as plasma volume expanders and smaller volumes of colloids are necessary in comparison to crystalloids for the same effect in plasma volume. Colloids can only be used for intravenous or intraosseous fluid therapy. Contraindications for the use of colloids include cardiac failure and kidney failure.

RESUSCITATION

Birds in hypovolemic shock should have their circulation normalized as soon as possible by expanding the intravascular volume. Intravenous or intraosseous fluid therapy is indicated. Colloids and/or crystalloids can be used.

FLUID RESUSCITATION PROTOCOL

Hypertonic saline (7.5%) 3 ml/kg + hetastarch 3 ml/kg as intravenous or intraosseous boluses administered over 10 minute, followed by boluses of crystalloids (e.g., lactated Ringer's solution) 10 ml/kg administered over 10 minutes. Boluses of crystalloids are repeated until clinical symptoms of shock have disappeared.

or

Boluses of crystalloids (e.g., lactated Ringer's solution) 10–20 ml/kg are administered over 2–10 minutes. Boluses of 10 ml/kg are repeated until clinical symptoms of shock have disappeared.

DOI: 10.1201/9781003308270-8

REHYDRATION & MAINTENANCE

Many sick birds are dehydrated by the time they present at the emergency consultation. In mild dehydration, there are often no noticeable abnormalities in the general examination. Clinical signs of severe dehydration include deep-set eyes, wrinkled skin, decreased skin turgor of the upper eyelid and the dorsal side of the feet, thick mucus in the oral cavity, delayed refilling time of the basilic vein (the vein on the ventral side of the wing near the elbow) and loss of body weight.

The refilling time of the basilic vein (**Fig. 6.1**) is determined by compressing the vein with the finger to push the blood out and observing how quickly the vessel is refilled after lifting the finger. When the blood visibly (and thus slowly) flows back instead of almost immediately returning, there is often severe dehydration. Make sure that the wing is kept in a neutral position during this test and is not lifted too far above the body, as the latter can lead to decreased perfusion.

Fig. 6.1 The basilic vein (arrow) is located on the ventral side of the wing close to the elbow (circle).

Blood tests showing increased packed cell volume (PCV), increased total protein and increased albumin level are indicative of dehydration. It should be noted that other factors can also influence these values, for example, an anemic bird with dehydration can have normal PCV.

Unfortunately, the degree of dehydration cannot usually be determined precisely in birds. A precise calculation of the amount of fluids needed for correction is therefore often not possible.

The fluid requirement of most birds consists of the maintenance requirement (approximately 50–100 ml/kg/24 hours) and the volume required to correct any dehydration and additional losses (due to polyuria, diarrhea and/or vomiting).

FLUID THERAPY (RULE OF THUMB)

The fluid requirement of most sick birds (maintenance and correction) can generally be estimated at 10% of the body weight per 24 hours for the first 3 days.

With the exception of patients who can receive a continuous infusion, this amount is usually divided over two to four doses per day. Fluid therapy can be oral and/or parenteral. In birds that receive fluid food, for example when formula is being fed through a feeding tube, the water in the formula can have a significant volume and also counts.

Example: A sick bird with a body weight of 100 g needs 10 ml of fluids on the first day. When this is divided over 3 doses per day, this is therefore 3.3 ml at a time.

In birds that are severely weakened, vomiting or possibly have a delayed emptying of the crop, oral fluid therapy is not initially chosen. In such cases parenteral fluid therapy is indicated.

ORAL FLUID THERAPY

Oral fluids are often an efficient method of fluid therapy in birds with mild dehydration and a functional gastrointestinal tract. Oral fluid therapy is probably less troublesome for the patient than parenteral fluid therapy (infusion).

Unfortunately, oral fluids are not effective when the functioning of the gastrointestinal tract has declined. Examples are situations with delayed emptying of the crop (see Crop stasis, p. 95) or vomiting/regurgitation (see Vomiting, p. 89). Fluid administered orally has no positive effect in these situations. In some cases, oral fluid therapy carries an excessive risk of aspiration (for example, in patients who are severely debilitated or have epileptic seizures).

Although administering fluids drop by drop into the beak with a syringe can be possible in some birds, oral administration of fluids is most efficient by administering the liquid directly into the crop with the aid of a feeding tube (see Appendix 5, p. 195).

Per 100 grams of body weight, usually 2–5 ml of moisture (2–5% of the body weight) can be slowly introduced into the crop. The lower limit is maintained for the first dose, the volume is gradually increased with subsequent doses. It is important that the temperature of the liquid is between 38–40°C (a bit below body temperature). This is to prevent regurgitation, hypothermia or burns of the crop. To avoid regurgitation and aspiration, care must be taken not to exert any external pressure on the crop during and immediately after insertion of the fluid.

PARENTERAL FLUID THERAPY

In cases where fluid therapy is necessary but oral fluid administration is not indicated or impossible, fluid administration is started by means of an infusion. See Appendix 2. An intravenous (**Fig. 6.2**) or intraosseous infusion (bolus or continuous flow) has the most immediate effect, but is technically more difficult than subcutaneous administration. In weakened patients, longer handling needed for the placement of an intravenous or intraosseous catheter by an inexperienced person may increase the risk of collapse. In most cases, administration of fluids by subcutaneous injection during emergency services by less-experienced veterinarians is the most responsible choice. However, in patients in shock, subcutaneous fluids may not be sufficiently absorbed, and an intravenous or intraosseous infusion is nevertheless indicated.

In parrots in particular, it is often a challenge to keep an intravenous catheter functional for any length of time, due to species behavior and anatomy. The veins are easily traumatized, sufficient immobilization of a wing can be unrewarding and some birds are very adept at destroying infusion systems (although most lethargic birds will not do this and accept the situation). Therefore, during an emergency consultation, it is usually decided to give a bolus via an intravenous catheter or hypodermic needle, instead of a continuous flow with the aid of an infusion pump.

As a bolus, usually 10–20 ml/kg can be administered over 2–10 minutes. For continuous infusion, rates of 3–10 ml/kg/hour are often suitable.

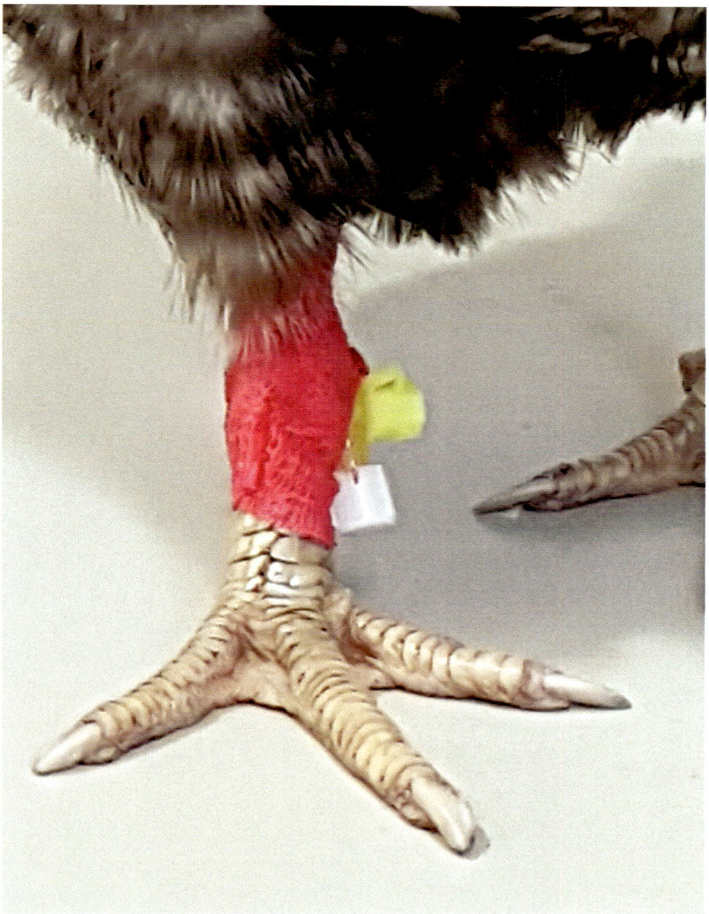

Fig. 6.2 IV catheter placed in medial metatarsal vein of a chicken.

In small birds, or birds with severely decreased blood pressure, administering an intravenous infusion might not be possible. In these situations, a needle or catheter can be placed in the medullary cavity of a bone for an intraosseous infusion. However, not all bones are suitable for intraosseous fluid administration. A number of hollow bones, such as the upper arm bone (humerus) and thigh bone (femur), are in direct connection with the air sacs, and administration of fluid into the medulla of these bones may result in the patient drowning! In most species, intraosseous infusion is safe in the proximal shinbone (tibiotarsus) and the distal ulna. The amount of fluid that can be administered and the rate of infusion are comparable to an intravenous infusion.

7 *Nutritional support*

Birds that don't eat enough run a great risk of developing nutritional and metabolic problems and dying. Many sick birds become anorectic. For this reason, assisted or forced feeding is started at an early stage in treatment (usually immediately). The fastest and often safest way is by delivering nutrition through a feeding tube introduced into the crop (see Appendix 5, p. 195). An additional advantage of tube feeding sick birds is that the commercially available formulas often have a better composition than the average diet of many pet birds, so that the patient also receives other essential nutrients in addition to energy and fluids.

The type of food used for tube feeding depends on the species and its natural diet. Herbivores (including the granivorous psittacines) and omnivores can be fed with formulas for juvenile psittacines or recovery formulas. Carnivores can be fed with commercially available liquid food for dogs and cats or a meat slurry made from pureed meat (intended for human consumption) mixed with physiological saline solution.

Note: In anorectic birds of prey, it is important to first provide oral fluids only to be able to assess whether the gastrointestinal tract has normal motility. Crop stasis (see p. 95) can quickly lead to potentially lethal sour crop (overgrowth of microorganisms in the crop) when meat is introduced and not passed to the stomach! When the passage of the liquid appears to be normal, increasingly more concentrated or thickened meat slurry can be administered.

TUBE FEEDING VOLUME

In sick anorectic adult birds that do not vomit, 20 ml/kg can, in most cases, be safely introduced initially. If this volume does not cause problems such as regurgitation, vomiting or crop stasis, the volume of the second dose can be increased to 30 ml/kg. If this does not cause any problems, the volume of the following doses can be gradually increased to 40 ml/kg, if necessary.

In birds that are nauseated or have an abnormality in the crop itself, the volume should be halved. This means that a smaller volume must be administered more often to still provide sufficient nutrition. As soon as the crop is empty, the next dose can be administered.

The amount of food introduced should be sufficient to prevent weight loss in birds with a healthy body condition and to gradually increase the body weight of thin birds. Measuring body weight (at least every morning before breakfast) is therefore very important.

DOI: 10.1201/9781003308270-9

Growing juvenile birds (up to a few weeks in small species and a few months in large species) have a relatively larger crop. Volumes can therefore be increased earlier, up to 10% of body weight. In young birds fed with a syringe at home, the usual feeding regimen can be maintained, provided there is no nausea or crop stasis (see p. 95). If a young bird is used to eating formula out of a syringe or from a spoon, food can be offered initially in the same way at the clinic. If food intake is too little this way, tube feeding is indicated. In very young birds, the next meal should be offered as soon as the crop is empty.

8 *Oxygen*

BIRDS with cyanosis, tachypnea or dyspnea (open beak breathing, forced breathing, abnormal breathing sounds/stridor) need additional oxygen.

Dyspnea in birds often has an extra-respiratory cause, such as a space-occupying lesion (e.g., an egg or enlarged organ) or free fluid in the body cavity, resulting in compression of the air sacs. Handling of these patients should be minimized and performed with as little pressure on the body as possible to avoid more severe dyspnea, or even death. In birds with dyspnea as a result of cardiovascular or respiratory diseases (e.g., lung, air sac or trachea disease), stress and handling should be minimized as well, as stress leads to more breathlessness.

Dyspneic birds should be placed in an oxygen cage. Birds that are severely short of breath on entry should be stabilized this way first before being handled for the physical examination.

DOI: 10.1201/9781003308270-10

9 *Analgesia and anesthesia*

ANALGESIA

Many emergency situations and other diseases lead to pain. Although birds don't show the same symptoms of pain as for example humans, dogs or cats, they do experience pain in the same way. Besides the fact that pain affects well-being adversely, it has many negative physical effects, including anorexia, inactivity and impaired wound healing. Whenever possible, pain should be suppressed or, even better, prevented.

In birds, analgesic drugs from several classes can be used as sole agents or combined for pain management.

NON-STEROIDAL ANTI-INFLAMMATORY DRUGS (NSAIDS)

NSAIDs relieve pain and reduce inflammatory reactions.

Possible adverse effects of NSAIDs include effects on the renal system, gastrointestinal system and coagulation. Contraindications include acute renal failure and gastrointestinal ulceration (signs of which can include melena, fresh blood in the droppings and vomiting).

Meloxicam is the most widely used NSAID in avian practice.

OPIOIDS

Opioids relieve pain by interacting with opioid receptors in the nervous system. Distribution of the different opioid receptors may vary between different species of birds. Possible adverse effects of opioids include cardiac and respiratory depression, decreased gastrointestinal motility and nausea.

Butorphanol is a short acting (1–3 hours) opioid with analgesic and mild sedative properties. Butorphanol can be administered IV, IM and IN (intranasal). Because of the short duration of action, butorphanol is useful as an analgesic for short painful episodes and procedures, for example surgery.

Tramadol has a longer duration of action than butorphanol. Tramadol can be administered PO in case of longer periods of moderate-to-severe pain that is refractory to NSAIDs.

LOCAL ANESTHETICS

Local anesthetics cause local analgesia without unconsciousness and are mainly used for analgesia during surgery in avian practice.

Lidocaine (preparations without epinephrine) is most commonly used for regional infiltration in birds.

DOI: 10.1201/9781003308270-11

GABAPENTIN

Gabapentin is an anti-convulsant drug that also relieves neuropathic pain.

SEDATION/ANESTHESIA

Sedation and general anesthesia is used in avian practice for procedures that either cause stress or require immobilization for maximum safety for patients, handlers or clinicians. In emergency situations, sedation or general anesthesia can, for example, be indicated for radiology, sample collection, application of bandages, fluid therapy and surgical procedures.

Like in other species, anesthesia in birds is complex. In this chapter, only a few agents will be discussed. Other agents can be used in avian practice, but risks of adverse reactions and even death are greater, especially when used by anesthetists with little experience in avian anesthesia.

MIDAZOLAM

Midazolam is a benzodiazepine used for sedation, muscle relaxation and decreasing anxiety. It is also an anti-epileptic drug. Midazolam can be administered IV, IM, IO and IN. Midazolam doesn't have analgesic properties.

ISOFLURANE/SEVOFLURANE

Isoflurane and sevoflurane are volatile anesthetic agents. Both agents cause loss of consciousness and are used for induction and maintenance of general anesthesia.

Isoflurane is the most widely used volatile anesthetic in avian practice. Induction generally occurs within minutes at a concentration of 3–5%. Induction is most frequently done with a face mask covering the beak and nares (preferably not the eyes to prevent irritation). For maintenance, 2% is generally effective.

Note: Painful procedures should not be performed with isoflurane or sevoflurane as the sole anesthetic agents! Analgesic agents should be included to provide adequate pain relief.

10 *Quick guide for stabilizing birds in case of severe dyspnea, debilitation and shock*

DUE TO their often small size and unique physical characteristics, birds can deteriorate rapidly in the event of illness and injury. Combined with their instinctive behavior of hiding symptoms, this frequently leads to birds arriving at the veterinary clinic in a severely debilitated or even moribund condition. These very critical patients frequently suffer from life-threatening hypothermia, hypovolemia, hypoxia and/or hypoglycemia caused by underlying disease or injury. Stabilizing birds in shock by managing hypothermia, hypovolemia, hypoxia and hypoglycemia as soon as possible is crucial for survival and frequently has priority above making a definitive diagnosis in emergency situations. This chapter is meant as a quick guide about how to act in situations with birds in severe debilitated state or shock. More information can be found in Part 2.

HYPOTHERMIA

The average body temperature of birds is 40°C, ranging from 39–43°C in most species. Signs of hypothermia in birds include lethargy and sitting still with fluffed feathers.

The ambient temperature must be raised to 30–32°C by means of an external heat source such as a heated incubator, heat lamp or heat mat. The optimal temperature varies by bird species, age and case presentation. Monitor patients placed in a heated environment for signs of hyperthermia (e.g., tachypnea, breathing with open beak and keeping the wings away from the body), which can be even more dangerous than hypothermia (see Hyperthermia, p. 195).

As hypothermia can be the result of anorexia, providing nutritional support (see Nutritional support, p. 33) is crucial as well.

HYPOVOLEMIA

Birds in hypovolemic shock should have their circulation normalized as soon as possible by expanding the intravascular volume. Intravenous or intraosseous fluid therapy is indicated. Colloids and/or crystalloids can be used.

Hypertonic saline (7.5%) 3 ml/kg + hetastarch 3 ml/kg as intravenous or intraosseous boluses administered over 10 minutes, followed by boluses of crystalloids (e.g., lactated Ringer's solution) 10 ml/kg administered over 10 minutes. Boluses of crystalloids are repeated until clinical symptoms of shock have disappeared.

DOI: 10.1201/9781003308270-12

or

Boluses of crystalloids (e.g., lactated Ringer's solution) 10–20 ml/kg administered over 2–10 minutes. Boluses of 10 ml/kg are repeated until clinical symptoms of shock have disappeared.

HYPOXIA

Handling of dyspneic birds should be minimized and done with as little pressure on the body as possible to prevent restricting respiratory movements.

Dyspneic birds should first be stabilized in an oxygen cage. Applying oxygen through a face mask can cause extra stress and an increased need for oxygen, thus having an adverse effect.

In case of obstruction of the trachea or syrinx, air sac intubation (see Appendix A6) is indicated. Signs of partial obstruction include severe dyspnea and stridor. When signs of dyspnea indicating an obstruction get worse or don't improve in 30 minutes despite oxygen therapy, intubation of a caudal air sac can be indicated as a temporary measure to create a clear airway.

HYPOGLYCEMIA

Hypoglycemia can be caused by anorexia or malnutrition, hepatic, kidney or intestinal diseases, hormonal diseases, sepsis and neoplasia. Signs of hypoglycemia include lethargy, ataxia and epileptic seizures.

Nutritional support and oral or parenteral fluid therapy with 5% glucose are indicated in addition to treating the underlying cause of hypoglycemia. Parenteral glucose supplementation can cause dangerous fluid and electrolyte disturbances and should be done with care.

Part 3: Specific emergency situations

11 *Leg band constriction*

Leg band constriction (**Fig. 11.1**) can initially be caused by a primary swelling of the leg (because of inflammation, trauma or tumor), accumulation of skin material and debris under the leg band, shifting of the leg band over the hock joint to a thicker part of the leg, deformity of the leg band caused by trauma or chewing and by using a leg band that is too small in growing birds. The constriction of the leg causes severely impaired blood flow out of the foot, leading to even more swelling. Serious complications can develop quickly, such as necrosis (death) of the foot, fractures of the tarsometatarsal bone (**Fig. 11.2**) and wounds in contact with the ring. The leg band itself may be hidden from view due to the position of the leg and covering feathers. Sometimes, inability to bear weight and/or lameness are the only clearly visible symptoms at home.

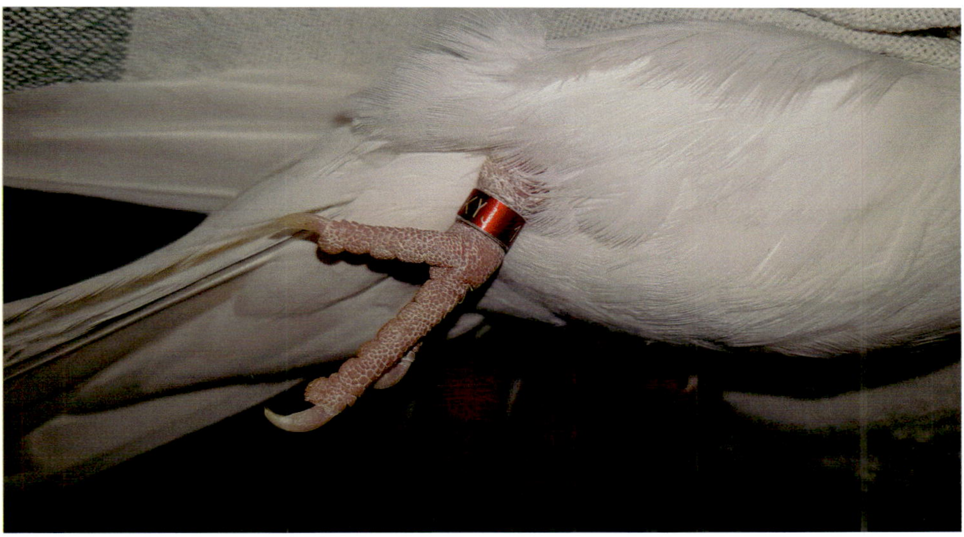

Fig. 11.1 Leg band constriction.

DOI: 10.1201/9781003308270-14

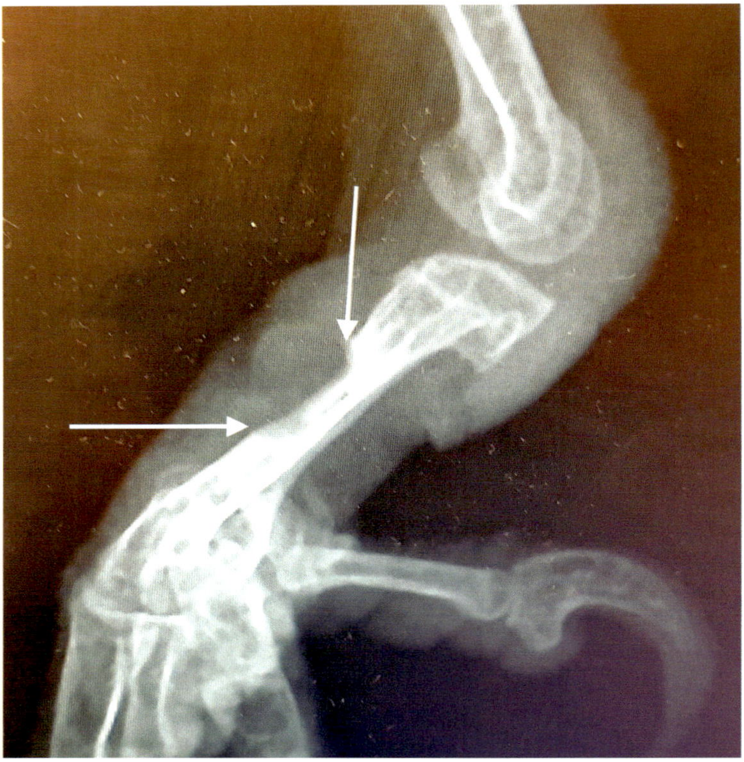

Fig. 11.2 X-ray of the foot of a macaw with chronic leg band compression. Note the lesion in the tarsometatarsal bone (between arrows). Leg band compression can lead to pathological fractures, which might not be obvious until after removal of the leg band.

ADVICE FOR FIRST AID AT HOME

Unfortunately there is nothing that can be done safely at home. Owners should not try to remove the leg band themselves. The risk of complications is too high.

EMERGENCY CARE BY A VETERINARY PROFESSIONAL

ANALGESIA

In case of swelling of the leg, analgesics (e.g., meloxicam and in severe cases butorphanol) are indicated.

REMOVAL OF THE LEG BAND

A leg band that is too tight should be removed as soon as possible. Trying to reduce the swelling in another way will be unrewarding. Cutting a leg band with scissors or pliers carries high risks. Sudden movement of the ring under pressure can cause very serious injury, including broken bones or complications leading to the loss of the foot or leg.

Cutting the leg band with a cooled high-speed drill/dental drill (**Fig. 11.3**) is, in the author's opinion, the best and safest option. Multi-tools with a rotating grinding disc can also be used, but have a higher risk of damaging the surrounding tissues, especially if the swelling protrudes above the leg band. When using a multi-tool, cooling the ring with a water spray/drop is essential for preventing heating of the ring and burning of the leg under the leg band. The patient being still is of utmost importance; this procedure can often best be done on birds under general anesthesia (e.g., with isoflurane/sevoflurane).

Tips for removal of the leg band:

- Start with a systemically acting painkiller (e.g., meloxicam) before the procedure.
- Prepare equipment in advance for treatment of potential bleeding or fracture stabilization. In severe cases, the soft tissues or bone may be so damaged that bleeding occurs spontaneously or a pathological fracture appears immediately after removal of the leg band. Before removing the leg band, inform the owner of these possible complications.
- Apply a layer of elastic bandage proximal and distal to the leg band to avoid iatrogenic soft tissue damage (especially when working with the multi-tool).
- When possible, place a thin metal object between the skin and leg band to protect the underlying tissues.
- Ensure continuous cooling of the leg band during cutting to avoid heating the leg band and burning the soft tissues under the ring.
- Make two cuts in the leg band to divide it into halves.
- Where possible, only cut on the lateral side of the leg. Cutting on the medial side of the metatarsus can lead to damage of the medial metatarsal vein. After the first cut on the lateral side, the leg band is rotated halfway round for the second cut, again on the lateral side. The second cut should only be done on the medial side when it is impossible to turn the leg band.

AFTERCARE

Any non-infected circular wound visible after removing the leg band (**Fig. 11.4**) can be treated topically, for example with honey-based or other antiseptic ointment twice a day. In case of deeper bacterial infections, antibiotics are indicated. Analgesics should be continued until the leg is no longer painful. In case of severe pain where the effect of NSAIDs is not sufficient for pain relief, opioids (e.g., butorphanol or tramadol) can be used in addition. Protective bandages can be applied to prevent contamination of the wound and to lower the risk of auto-mutilation.

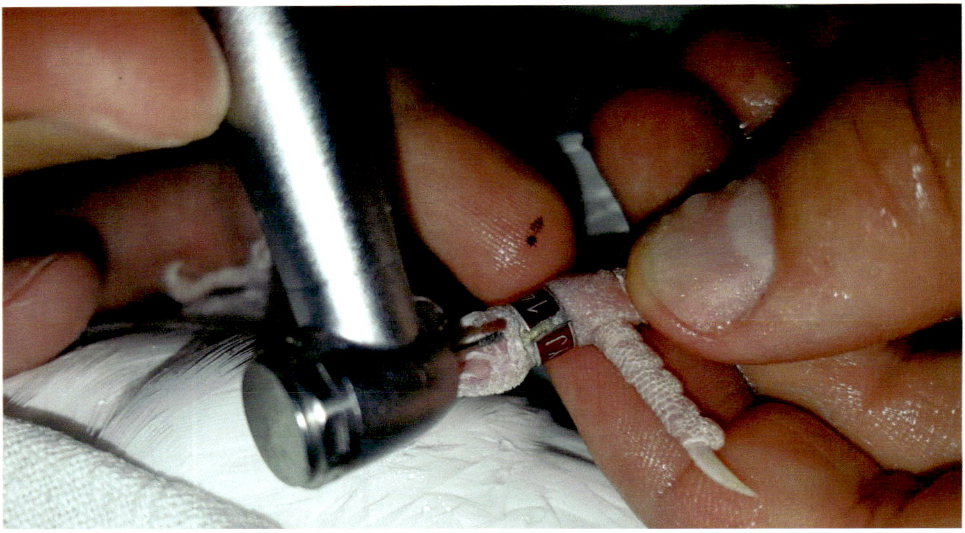

Fig. 11.3 Cutting the leg band with a cooled high-speed drill.

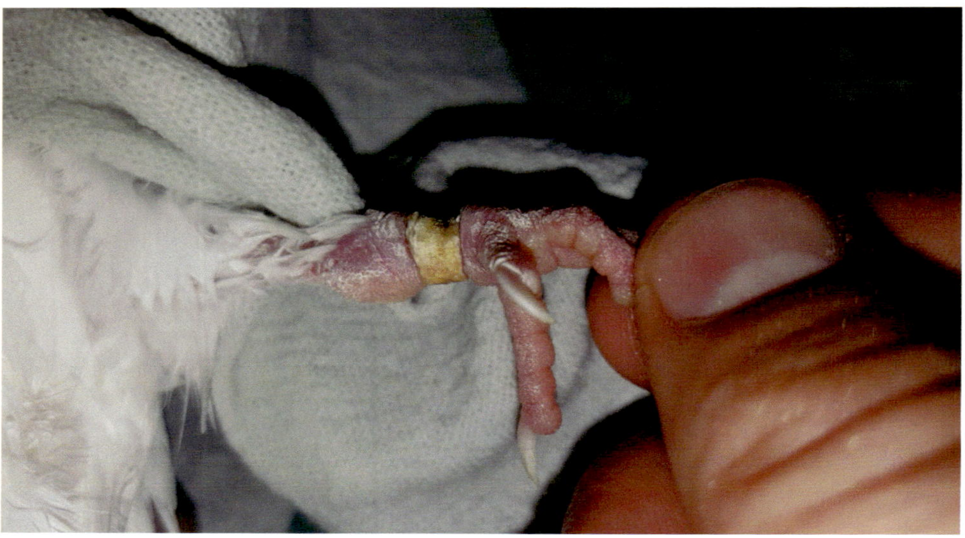

Fig. 11.4 Circular wound with debris remaining after removal of the leg band.

12 *Bleeding pin feather*

PIN OR blood feathers are normal new developing feathers with good blood supply. Damage to pin feathers (**Fig. 12.1**) can cause significant blood loss.

Fig. 12.1 Bleeding pin feather.

ADVICE FOR FIRST AID AT HOME

The bleeding end of the feather can be pinched firmly for several minutes. This can be done using fingers or pliers. A small amount of styptic powder, iron chloride or a silver nitrate stick can be used to stimulate blood clotting. Oral uptake of the chemicals by the bird should be prevented. If these chemicals are not available, a small amount of flour or cornstarch can also be used to stimulate blood clotting. Ligating or tying a tight knot using a string, thread or fishing line at the end of the bleeding feather can also be done, if possible.

In the situation where bleeding doesn't stop and it is not possible to get to a veterinarian on time, owners can pull out the bleeding feather themselves as a final

DOI: 10.1201/9781003308270-15

option. For this, the wing or tail must be held firmly at the base of the feather. Then the damaged feather is held firmly at the base with pliers and pulled out. Resulting minor bleeding from the feather follicle can be stopped by applying gentle pressure. Chemicals should not be used to stop bleeding from the feather follicle.

Note: Pulling out feathers is a very painful procedure that can lead to complications and should, when possible, be done by a veterinarian with sufficient analgesia.

* * *

EMERGENCY CARE BY A VETERINARY PROFESSIONAL

As described previously, an attempt can be made to stop the bleeding by pinching the feather for several minutes or by using styptic powder, iron chloride or silver nitrate.

Ligation of the pin feather should be attempted, and the feather allowed to molt out normally, if possible, to avoid trauma to the follicle and potential future complications in feather growth. Ligation should be performed proximal to any breaks or weak points in the feather shaft, and the feather trimmed distal to the ligation.

If the bleeding does not stop, the feather can be pulled out. Provide adequate analgesia if time and patient condition permit, e.g., butorphanol, meloxicam and/or lidocaine (injection at the base of the feather follicle or drop on the damaged feather shaft). While the wing or tail is properly restrained at the base of the feather (to prevent iatrogenic fractures or other trauma), the feather can be grasped with a needle holder or mosquito hemostats and be pulled out. In contrast to full-grown flight feathers, this is usually successful with relatively little force. If minor bleeding occurs from the feather follicle, this can be stopped by applying gentle pressure. Cauterization or the use of chemicals such as iron chloride or silver nitrate to stimulate blood clotting can cause permanent damage to the follicle and lead to complications.

Note: Pulling out feathers can cause permanent damage to the feather follicles. Although it is uncommon, the result could be that the new feathers become abnormal (dystrophic) or that no new feathers grow at all. Pulling out a flight feather should be seen as a last resort, particularly in birds of prey, where imperfect flying can lead to reduced survival in the wild.

FLUID THERAPY

Fluid therapy (see p. 27) is indicated in case of significant blood loss (more than 1% of body weight).

13 *Hyperthermia*

THE AVERAGE body temperature of birds is 40°C, ranging from 39°C to 43°C in most species. Hyperthermia usually occurs due to an excessively warm ambient temperature, but can also be induced by stress. Even tropical birds can get hyperthermic on hot days if they don't have enough shelter.

Birds with hyperthermia often breathe with their beaks open and usually keep the wings away from the body. If not treated on time, hyperthermia leads to death.

ADVICE FOR FIRST AID AT HOME

Overheated birds should immediately be placed in a cooler environment. The feet can repeatedly be moistened with fresh water and a water spray can be used to cool down the patient, but the plumage of the bird should not be soaked.

EMERGENCY CARE BY A VETERINARY PROFESSIONAL

FLUID THERAPY

Fluid therapy through subcutaneous, intravenous or intraosseous infusion (see Fluid therapy, p. 27) is indicated, but excessive stress and further overheating due to handling should be avoided.

COOLING

The patient is placed in a calm and cool environment. Room temperature is usually cool enough. Spraying fresh water on the legs and rest of the bird can contribute to accelerated cooling. The plumage should not be soaked. Alcohol applied to the feet can be even more effective than water, but is a toxic substance and should be used with care.

Note: Active cooling can quickly lead to hypothermia in birds. As soon as the symptoms of hyperthermia (abnormal breathing and posture) have disappeared, the additional measures should be stopped to prevent hypothermia. Signs of hypothermia in birds include lethargy and sitting still with fluffed feathers (see Hypothermia, p. 39).

DOI: 10.1201/9781003308270-16

14 *Bleeding nail or beak tip*

TRAUMA to the nails and tip of the beak (**Fig. 14.1**) can lead to significant blood loss, especially when birds continuously remove the forming blood clot.

Fig. 14.1 Bleeding tip of the beak.

ADVICE FOR FIRST AID AT HOME

Attempts can be made to stop the bleeding by applying gentle pressure with a sterile gauze to the bleeding area. In the case of a nail, gently squeezing the toe at the base of the nail can reduce blood flow. Styptic powder or silver nitrate sticks can be used to stimulate blood clotting. The application of a small amount of flour or cornstarch can also contribute to the formation of a blood clot.

It is important to prevent the patient from removing the clot with the beak. Protection of the nails can be accomplished by applying a bandage or tape around the damaged nail and toe.

DOI: 10.1201/9781003308270-17

EMERGENCY CARE BY A VETERINARY PROFESSIONAL

FLUID THERAPY

In case of significant blood loss (more than 1% of body weight), fluid therapy (see p. 27) is indicated due to the risk of hypovolemia.

ANALGESIA

Treatment with analgesic drugs for adequate pain relief must be started immediately in case of trauma to the beak. In birds with hypovolemia/hypotension due to significant blood loss (more than 1% of body weight), it is advisable not to start immediately with an NSAID, as this might increase the risk of acute kidney injury. Butorphanol can be used in the acute phase. After stabilization, NSAIDs (e.g., meloxicam) can be used instead.

MANAGEMENT

Initially, the same techniques as described previously can also be applied during the emergency consultation. By professionally handling/restraining or sedating a patient (e.g., by means of isoflurane/sevoflurane or midazolam combined with butorphanol), better results can be achieved in the clinic with the same techniques than at home. Non-sedated birds should be held until the bleeding has stopped. In some cases of bleeding of the tip of the upper beak, it can be useful to keep the beak closed with the fingers to prevent birds from continuously removing the newly formed blood clot with the lower beak. Of course, this should only be done in birds that are able to breathe freely through the nose.

If the bleeding does not stop in this way, cauterization can be considered. Given the pain of this procedure, cauterization should only be done with adequate analgesia (e.g., by means of butorphanol and meloxicam or topical lidocaine). Thermal damage to the bone must of course be minimized. Cauterization is therefore also the last remedy to be used, especially in case of trauma to the beak!

Sealing the defect of a nail or beak can help prevent repeated bleeding and reduce pain afterward. For the beak, this can be done by applying a thin layer of epoxy resin or tissue adhesive over the defect. This is done under general anesthesia. In the case of a nail where the horn is completely absent, a thin layer of tissue adhesive can be applied, followed by sprinkling it with a small amount of baking soda to add some volume to the protective layer. This step can be repeated several times to create a sufficiently thick layer.

AFTERCARE AND ANALGESIA

Beak trauma in particular can be painful for a long period of time and can lead to anorexia in addition to an impairment of well-being. The use of NSAIDs (e.g., meloxicam) is therefore indicated. In the case of beak trauma, it is sometimes necessary to offer soft food or to support the patient with tube feeding (see Nutritional support, p. 33) until food uptake is sufficient again. Most birds with trauma to the tip of the beak eat normally when treated with a painkiller in the form of an NSAID.

15 *Perforating (bite) trauma of the beak*

PERFORATING beak trauma is usually the result of a bite from another psittacine bird. However, it can also be caused by other trauma, such as an accident, dog bite, certain toys or metal objects. With deep beak trauma (**Fig. 15.1**), there is a high risk of contamination and infection of bone tissue. Serious complications can develop if adequate action is not taken in time.

Fig. 15.1 Severe trauma to the beak of a cockatoo.

ADVICE FOR FIRST AID AT HOME

Aseptic gloves should be worn when dealing with open wounds. A fresh bite wound on the beak may be rinsed with sterile saline solution, cooled boiled water or tap water if necessary. Foreign material such as feathers or bedding material present in the wound should be removed. In case of injuries that initially bled, care should be taken not to remove blood clots, because this can lead to more blood loss.

* * *

DOI: 10.1201/9781003308270-18

EMERGENCY CARE BY A VETERINARY PROFESSIONAL

Deep injuries to the beak should be treated as open fractures. There is a high risk of bacterial infection of damaged bone tissue.

ANALGESIA

Deep trauma to the beak causes severe pain, and because of that, treatment with analgesic drugs (e.g., meloxicam and butorphanol or tramadol) for adequate pain relief must be started immediately. In most cases, treatment with an NSAID should be continued for a few days to 1 week. In some cases, however, it is necessary to give painkillers for longer periods or to give tramadol in addition.

FLUID THERAPY

In case of significant blood loss (more than 1% of body weight), fluid therapy (see p. 27) is indicated due to the risk of hypovolemia.

WOUND TREATMENT

Any foreign material such as feathers, bedding material or dirt should be removed. If the wound is already older, samples for bacteriological culturing and susceptibility testing are taken. Completely loosened parts of horn (keratin) must be removed initially; in some cases, big loose fragments of horn can be used in a later stage, when a protective layer is made with epoxy resin or tissue adhesive to cover bigger defects. Inverted but still-attached parts of horn should be put back gently into the normal position using tweezers or mosquito forceps/hemostats. The wound must then be rinsed thoroughly with physiological saline solution and disinfected (with povidone iodine 1% or chlorhexidine 0.05%); see **Fig. 15.2**.

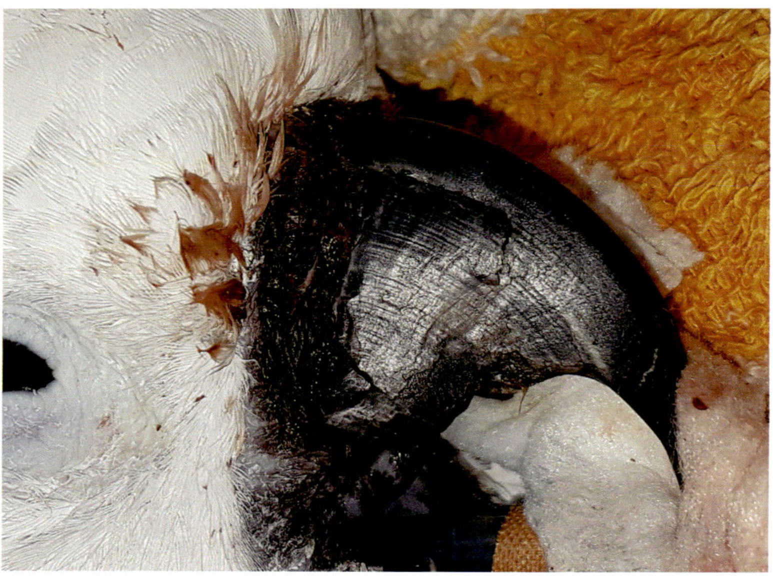

Fig. 15.2 Beak after cleaning and repositioning of horn fragments.

After drying, the defect can be covered with a protective layer of epoxy resin (**Fig. 15.3**).

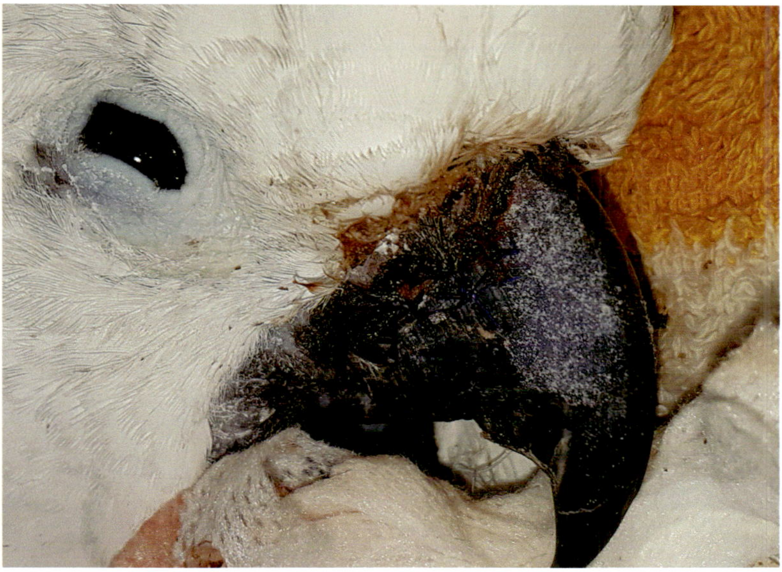

Fig. 15.3 Defect covered with protective layer.

ANTIBIOTICS

Due to the high risk of infection, treatment with a broad-spectrum antibiotic (e.g., amoxicillin/clavulanic acid) should be started immediately. In the case of bacteriological culturing and susceptibility testing, treatment with antibiotics is started pending culture results.

NUTRITIONAL SUPPORT

Beak trauma can lead to anorexia. It may be necessary to offer soft food or to support the patient with tube feeding (see Nutritional support, p. 55) until food uptake is sufficient again.

Treatment by an avian specialist is advisable in case of large, deep defects and management of possible permanent abnormalities of the beak, which can develop as result of severe trauma.

16 *Lacerations and cuts*

ADVICE FOR FIRST AID AT HOME

Aseptic gloves should be worn when dealing with wounds.

In bleeding wounds, gentle pressure should be applied to the bleeding areas, preferably with sterile gauze or clean towels. When the bleeding doesn't stop, gentle pressure should be applied continuously on the wound on the way to the veterinarian to avoid excessive blood loss.

Fresh, non-bleeding wounds may be rinsed with physiological saline solution and then disinfected once with dilute povidone iodine (diluted with physiological saline solution or lukewarm tap water until the solution has the color of light tea).

Wounds of the feet can be covered with a bandage or tape for protection and against soiling.

* * *

EMERGENCY CARE BY A VETERINARY PROFESSIONAL

FLUID THERAPY

In case of significant blood loss (more than 1% of body weight), immediate fluid therapy (see p. 27) is indicated due to the risk of hypovolemia. In birds in shock, intravenous or intraosseous administration is indicated. In stable patients with mild blood loss, subcutaneous fluid therapy is effective.

ANALGESIA

Treatment with analgesic drugs for adequate pain relief must be started immediately. In birds with hypovolemia/hypotension due to significant blood loss (more than 1% of body weight), it is advisable not to start immediately with an NSAID, as this might increase the risk of acute kidney injury. Local lidocaine and butorphanol can be used in the acute phase. After stabilization, NSAIDs (e.g., meloxicam) can be used instead of or in addition to the opioids (e.g., tramadol).

In most cases, treatment with analgesic drugs should be continued for a few days to 1 week. In some cases, however, it is necessary to give painkillers for longer periods.

WOUND MANAGEMENT

In older wounds, samples for bacteriological culturing and susceptibility testing are taken.

DOI: 10.1201/9781003308270-19

Surgery

Surgery should be done after wound cleaning and disinfection (with povidone iodine 1% or chlorhexidine 0.05%) or under general anesthesia (e.g., with isoflurane/sevoflurane) with adequate analgesia.

Bleeding from damaged bigger blood vessels must be stopped by ligation or cauterization. Diffuse bleeding can be stopped by applying gentle pressure to the bleeding surface.

Bigger, relatively fresh wounds (**Fig. 16.1**) should be closed with sutures for primary wound healing (**Fig. 16.2**). As non-surgical wounds can be contaminated or infected, monofilament suture material is preferred over multifilament suture material.

Note: Wounds older than 6 hours may require debridement to remove devitalized or infected tissues and to create fresh wound edges before closure.

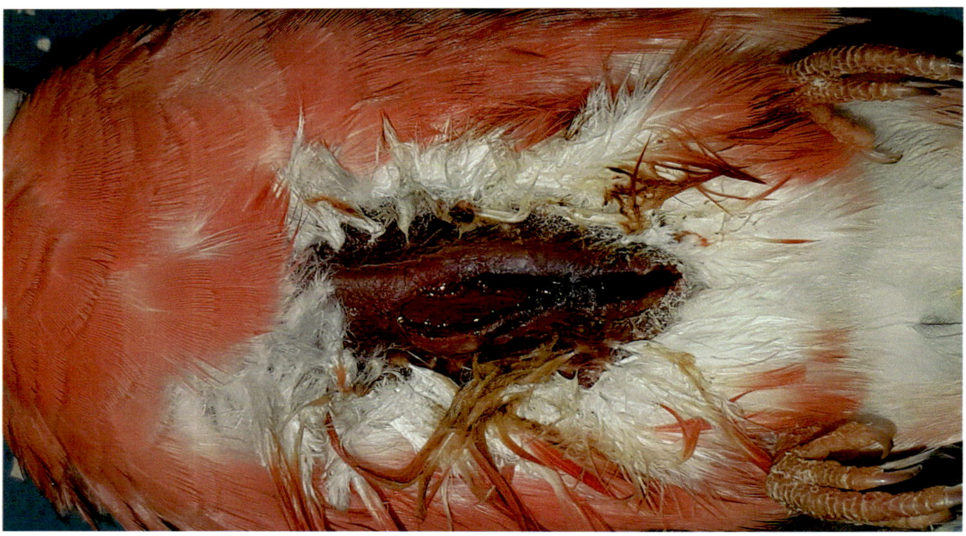

Fig. 16.1 Deep wound over the keel of a Bourke's parrot caused by repeated falling after losing all flight feathers.

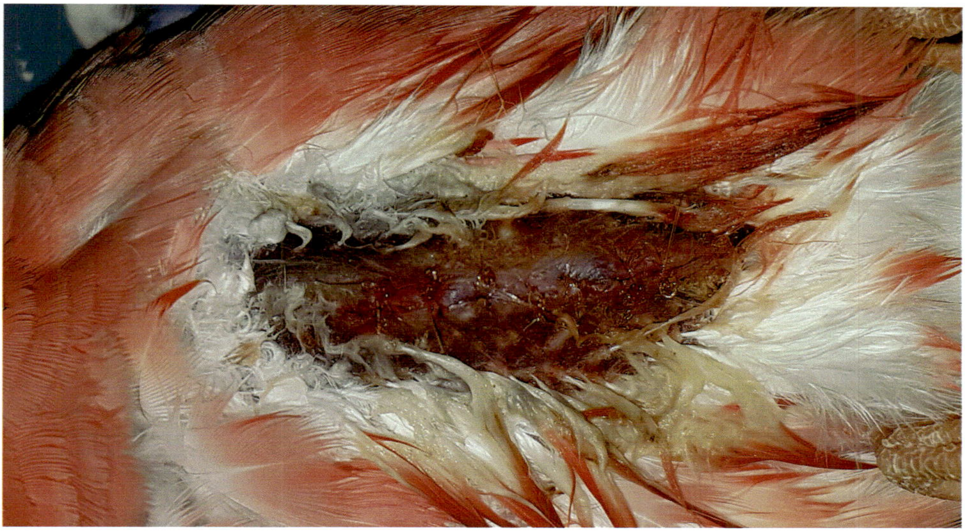

Fig. 16.2 After cleaning and disinfecting, the wound was closed with simple interrupted sutures.

Conservative treatment

Old wounds, large wounds with parts of skin missing and infected wounds can sometimes best be left open to heal by second intention. Repeated debridement (removal of necrotic tissues and debris) and treatment of infections (preferably based on the results of bacteriological culturing and susceptibility testing) are indicated until the wound bed consists of a healthy granulation tissue.

Optimal circumstances for wound healing can be created by applying wound dressings (to protect against contamination, absorb exudate and promote granulation) secured with bandages or stitches and/or topical preferably water-based preparations (e.g., silver sulfadiazine cream) or medical honey wound gel. Wound dressings should be changed at least once every 24 hours until the wound bed consists of a healthy granulation tissue.

Immobilizing injured body parts can help to prevent complications in wound healing caused by too much movement (see Appendix 8, p. 209).

ANTIBIOTICS

Antibiotics are indicated in heavily contaminated or infected wounds and in wounds of immunocompromised birds. The latter category includes very young birds, birds with chronic wounds, birds with comorbidity and birds suffering from malnutrition. Antibiotics are preferably based on the results of bacteriological culturing and susceptibility testing. In case of heavy contamination in fresh wounds or severe infections, treatment with broad-spectrum antibiotics (e.g., amoxicillin/clavulanic acid) might be started pending test results.

17 *Bite wound or deep wound caused by claws*

BITE WOUNDS (especially those caused by mammals like dogs, cats, rats, etc.) or deep wounds caused by claws are associated with a high risk of complications due to the risk of wound infection and damage to deeper structures, even when the skin wounds are small!

ADVICE FOR FIRST AID AT HOME

Aseptic gloves should be worn when dealing with wounds.

The wound should be rinsed immediately with physiological saline solution, cooled boiled water or tap water. Except for head injuries (to avoid contact of the disinfectant with the eyes), wounds should be disinfected once with dilute povidone iodine (diluted with physiological saline solution or lukewarm tap water until the solution has the color of light tea).

EMERGENCY CARE BY A VETERINARY PROFESSIONAL

ANALGESIA

Deep trauma causes severe pain, and because of that, adequate pain relief must be started immediately with, for example, butorphanol, meloxicam and/or lidocaine. With significant blood loss (more than 1% of body weight), it is advisable not to start immediately with an NSAID. Local lidocaine and butorphanol can be used in the acute phase. After stabilization with fluid therapy, NSAIDs (e.g., meloxicam) can be used instead of or in addition to the opioids (e.g., tramadol).

FLUID THERAPY

Fluid therapy (see p. 27) is indicated to counteract hypovolemia in cases of significant blood loss (more than 1% of body weight) or large hematomas.

WOUND MANAGEMENT

Samples for bacteriological culturing and susceptibility testing are taken. Wounds should be closed with sutures (if possible, without causing significant tension) after wound cleaning and disinfection (with povidone iodine 1% or chlorhexidine 0.05%). Wounds perforating the body wall into the coelom or air sacs should not be flushed because of the risk of the entry of fluids into the respiratory system.

Note: Wounds older than 6–12 hours may require debridement to remove devitalized or infected tissues.

DOI: 10.1201/9781003308270-20

Surgical drains should not be placed, as they are of no use in birds and only lead to complications.

ANTIBIOTICS

In cases of perforating bite wounds or deep wounds caused by claws/talons, the risk of bacterial infection is very high. Large numbers of bacteria can be introduced into severely traumatized and sometimes devitalized tissue. Starting treatment with a broad-spectrum antibiotic (e.g., amoxicillin/clavulanic acid) as soon as possible is often crucial. Antibiotic treatment should be started pending culture results.

18 *Self-mutilation*

SELF-MUTILATION or auto-mutilation is to direct injury by the biting of one's own skin or deeper tissues. In case of self-mutilation, immediate action is essential to prevent escalation, leading to further tissue damage and blood loss. Birds that self-mutilate should therefore be treated urgently by a veterinarian.

Feather-damaging behavior is not considered self-mutilation and is not an emergency. Due to the complexity of this problem, birds with feather-damaging behavior are better examined and treated by an avian veterinarian within opening hours.

ADVICE FOR FIRST AID AT HOME

Aseptic gloves should be worn when dealing with wounds.

Owners should try to stop the behavior by distracting the bird, letting it eat, putting it in the dark or holding it (at least in case of blood loss). In case of auto-mutilation of the feet, owners can apply a bandage around the injured body part.

* * *

EMERGENCY CARE BY A VETERINARY PROFESSIONAL

Note: Self-mutilation can have many different physical and psychological causes. During the emergency consultation, the patient must be stabilized and further escalation prevented. Determining the cause of the problem and its treatment can then be done at a later stage.

FLUID THERAPY

In case of significant blood loss (more than 1% of body weight), immediate fluid therapy (see p. 27) is indicated due to the risk of hypovolemia.

ANALGESIA

Treatment with analgesic drugs for adequate pain relief must be started immediately. In birds with hypovolemia/hypotension due to significant blood loss (more than 1% of body weight), it is advisable not to start immediately with an NSAID, as this might increase the risk of acute kidney injury. Local lidocaine and butorphanol can be used in the acute phase. After stabilization, NSAIDs (e.g., meloxicam) can be used instead of or in addition to opioids (e.g., tramadol).

DOI: 10.1201/9781003308270-21

PREVENTING AUTO-MUTILATION

Making it physically impossible to reach the site of damage with the beak is necessary to prevent further damage. This is, in most cases, most reliably accomplished by applying a neck collar (tube, ball and/or Elizabethan collar, **Fig. 18.1**). In cases of injuries to just the torso, unlike injuries to the legs or wings, a protective bodysuit can also be constructed out of a sock. However, bodysuits are hard to apply and don't stop every bird from biting. Thus, a neck collar is preferred over a bodysuit in an emergency situation. Self-mutilation of the legs can sometimes be prevented by applying a protective bandage, consisting of a layer of padding or gauze compress, elastic self-adhesive bandage and layers of tape. Birds biting the bandage or redirecting the self-mutilation to another body part may need a neck collar in addition to the bandages.

Fig. 18.1 A neck collar to prevent auto-mutilation of the leg.

Note: Neck collars can cause severe stress, depression and anorexia. Birds with recently applied neck collars should be hospitalized for monitoring by veterinary professionals until the collar has been fully accepted.

WOUND TREATMENT

The wound itself can be treated as described for lacerations and cuts (see, p. 57).

Further diagnostics and treatment are indicated after stabilization.

19 *Burn injuries*

Burn injuries can result from contact with chemicals or a hot object, liquid or gas.

ADVICE FOR FIRST AID AT HOME

Any part of the body that has come into contact with a hot liquid, object, gas or open flame should ideally be cooled immediately. Feet can be cooled with running lukewarm tap water or by placing the bird in a foot bath. This is an essential part of the treatment of burn injuries to the feet and should already be done at home. Acute burn injuries of the back can be cooled with running lukewarm tap water for a short period as well, but care must be taken to prevent undercooling of the bird. Owners should not soak the feathers and only use the water for a few minutes. In burn injuries to the head, cooling with running lukewarm tap water is usually not possible in a safe manner. A water spray applied repeatedly can be used instead.

* * *

EMERGENCY CARE BY A VETERINARY PROFESSIONAL

After cooling, pain relief and prevention of wound infection are essential to minimize risks of complications and harm to well-being.

ANALGESIA

Burn injuries are very painful and analgesia is indicated. Because severe burn injuries can lead to shock, opioids (e.g., butorphanol or tramadol) are preferred to NSAIDs in the acute phase. After stabilization with fluid therapy, NSAIDs (e.g., meloxicam) can be used instead of or in addition to opioids.

FLUID THERAPY

Fluid therapy (see p. 27) is indicated.

PREVENTION/TREATMENT OF INFECTION AND WOUND TREATMENT

In older wounds, samples for bacteriological culturing and susceptibility testing are taken.

Open wounds can be treated with an anti-microbial cream (e.g., silver sulfadiazine cream or an antibiotic cream without corticosteroids) twice a day. Application of the cream should be done with gloves on to avoid infection with bacteria from the hands of the practitioner.

DOI: 10.1201/9781003308270-22

Legs with burn injuries with or without visible skin lesions are best wrapped in a bandage. Open wounds must first be covered with a layer that does not stick to the wound (e.g., an ointment-impregnated sterile gauze) and then protected with an adhesive bandage. The bandages protect against contamination with feces or environmental bacteria and auto-mutilation. Burn injuries on the back can be covered with a body wrap bandage, but it is less necessary, because there is little or no contact with feces or bedding. Burn injuries on the head can, in most cases, not be covered with a protective bandage.

After 3 days, the skin condition should be assessed. Necrotic tissue should be removed and wounds treated.

* * *

CHEMICAL BURN INJURIES

Certain chemicals, such as bleach, can also cause burn injuries. In case of contact with chemical substances, the priority is to remove the chemicals from the skin as soon as possible. Avoid transferring the chemicals to any other area of the skin, mucous membranes or eyes. The skin should then be rinsed thoroughly with lukewarm tap water.

After removing the caustic substance, the wound can be treated as a regular burn.

Note: Although it may take several weeks, even large areas of burn injuries to the skin can, in many cases, heal well by second intention with conservative therapy, provided that colonization by microorganisms is prevented.

20 *Contact with glue from rodent or insect trap*

GLUE FROM a rodent or insect glue trap on feathers or the body can have serious consequences.

ADVICE FOR FIRST AID AT HOME

Owners should put on gloves to prevent the glue from getting on their own skin. Sticky areas can be covered with tissues to prevent further spreading of glue over larger parts of the body. If the bird is glued to an object, the bird must be released very carefully. First, owners can try to pull the feathers off the sticky surface one by one. No attempts should be made to try to free the bird at once by pulling it away from the sticky surface. Besides being painful, this can cause severe skin trauma and fractures of the legs and wings. When attempts to free the bird by pulling the feathers from the sticky surface are unsuccessful and the bird is not in severe distress caused by a very uncomfortable position (such as a limb in an unnatural position), owners can take both to the clinic. When the object is too large to take to the clinic or the bird is in severe distress caused by a very uncomfortable position, the glued feathers can be cut very carefully with scissors. Of course, the skin should not be cut, which can be challenging. Immediately cover released sticky body parts with tissue paper.

Owners should try to prevent the bird from trying to remove the glue with the beak itself by holding or distracting it, even on the way to the veterinarian.

When only a few feathers are smeared with glue, an attempt can be made to remove the glue at home. Unfortunately, there are different types of glue and there is no one standard way to remove them all. Chemicals that can dissolve glue are often very toxic and should therefore not be used. Many glues that are used against rodents or insect pests dissolve in oil. Those glues can be removed from feathers by rubbing the feather with non-toxic oil, for example olive oil, salad oil, sunflower oil or peanut butter. Although this oil is not toxic, birds should not ingest this oil either, especially not after dissolving the glue. Contamination of unaffected feathers with oil should be prevented. Baby wipes can also help to remove certain glues. After dissolving the glue, feathers must first be cleaned by wiping with a clean towel. Residual oil can then be carefully washed off with lukewarm, highly diluted washing-up liquid and then rinsed with lukewarm water.

Note: Excessive washing can cause a decrease in body temperature (hypothermia), which can be dangerous. Birds with lots of feathers or big areas of the body covered with glue should be treated by a veterinarian. Birds with glue that cannot be dissolved in oil or removed with baby wipes should be treated by a veterinarian as well.

* * *

DOI: 10.1201/9781003308270-23

EMERGENCY CARE BY A VETERINARY PROFESSIONAL

Initially, the procedures described previously can be attempted in the clinic. General anesthesia (e.g., with isoflurane/sevoflurane) can facilitate the procedure. If this does not work, cleaning the feathers or skin with propylene glycol can be attempted.

In severe cases, smeared feathers can be removed under general anesthesia (e.g., isoflurane/sevoflurane or midazolam combined with butorphanol) with adequate analgesia (e.g., butorphanol and meloxicam). Pulling feathers is preferred to cutting, because pulling feathers results in faster regrowth, while after cutting, new feathers will only appear during the next molting period.

21 *Oil contamination*

OIL SPILLS form a big danger to aquatic birds. Oil covering the feathers inhibits normal thermoregulation, impairs waterproofing and causes an inability to fly. Ingestion of oil can lead to severe gastrointestinal problems (e.g., anorexia, vomiting/regurgitation, diarrhea, malabsorption) and, depending on the specific type of oil, can cause systemic intoxications leading to anemia, kidney or liver damage. Aspiration of oil leads to dyspnea.

Many untreated birds die because of emaciation, hypothermia or drowning.

ADVICE FOR FIRST AID AT HOME

Oiled birds should only be held with gloves on or with towels. Oiled birds should be placed in a warm and dark environment. No attempts should be made to remove the oil. Oiled birds found outside must be seen by a veterinary professional or specialized rescue center as soon as possible.

EMERGENCY CARE BY A VETERINARY PROFESSIONAL

Oil should be removed from the oral cavity, choana and pharynx with cotton tips/swabs to clear the airways. The eyes can be cleaned with cotton tips/swabs and by flushing with NaCl.

Oiled birds are frequently severely debilitated and in shock because of hypothermia, hypoglycemia and dehydration. Oiled birds must be stabilized first (see Chapter 10, p. 39) before attempts are made to remove the oil from the feathers.

After general stabilization (this might take days), the oil should be removed from the feathers. This is done by washing oiled birds with flowing warm water (40°C) with 1–2% diluted liquid dishwashing soap (such as Dawn). The head, especially the area around the eyes, can be cleaned using soft toothbrushes or cotton swabs. Birds should be monitored carefully during washing, especially for signs of hypothermia. Washing should be stopped when birds become unstable again and must only be continued after fully stabilizing the patient. In the end, all oil should be removed, but it might take several washing sessions.

DOI: 10.1201/9781003308270-24

22 *Intoxications*

UNFORTUNATELY, intoxications are quite common in birds. Intoxication can occur at home by ingesting, inhaling or touching toxic substances. Psittacines in particular like to bite on all kinds of things and therefore run a relatively high risk when they are in a room with toxic substances, such as batteries, poisonous plants, chocolate, avocado, alcohol, cleaning products, tobacco, drugs, or lead. All birds are very susceptible to intoxication by inhaling toxic gases and vapors.

Many countries have poisoning information centers that can be contacted by professionals for information about specific intoxications. Examples are the American Society for the Prevention of Cruelty to Animals (ASPCA) Animal Poison Control Center at ☎(888) 4264435 in the USA and Veterinary Poisons Information Service at ☎02073055055 in the UK.

In this chapter, the following intoxications will be discussed:

1. Inhalation intoxication
2. Contact of the skin or eyes with a toxic substance
3. Oral intoxication
   - Lead poisoning
   - Poisonous plants
   - Corrosive toxins
   - Other toxins

INHALATION INTOXICATION

Inhalation of toxic fumes, gases or smoke can poison birds through the respiratory tract. Examples include exposure to the toxin polytetrafluoroethylene (PTFE) (a toxic gas that is released when this non-stick coating on cookware is overheated), carbon monoxide, smoke, cooking fumes, and certain cleaning agents/solvents. Most inhaled toxins cause damage to the respiratory tract, but problems in other organs are possible as well with certain toxins.

Symptoms of inhalation intoxication depend on which toxin is inhaled. Dyspnea (shortness of breath) is seen most commonly as a result of problems in the airways, including fluid accumulation in the lung tissue (pulmonary edema), narrowing of the airways (bronchoconstriction), fluid discharge in the airways (exudation), or mucous membrane swelling. Secondary infection can cause additional problems.

ADVICE FOR FIRST AID AT HOME

Birds must immediately be removed from the area where exposure took place. Owners should report the substance the bird has been exposed to so the

DOI: 10.1201/9781003308270-25

veterinarian can collect information about the specific intoxication and start appropriate treatment as soon as the bird arrives for the emergency consultation.

* * *

EMERGENCY CARE BY A VETERINARY PROFESSIONAL

Birds with signs of dyspnea should be placed in an oxygen cage immediately.

If necessary, information about the specific poison should be obtained by calling a poisoning information center.

General stabilization

Sick patients are stabilized using heat, nutrition, fluid therapy, oxygen (see General stabilization of sick birds, p. 23).

When possible, a specific treatment against the intoxication is started.

Symptomatic treatment

Anti-inflammatory drugs (for example, meloxicam) are indicated in case of inflammation and irritation of the airways. Furosemide is indicated in case of pulmonary edema. Salbutamol is indicated in case of bronchoconstriction.

After inhalation of an agent that causes damage to the airways, broad-spectrum antibiotic treatment (e.g., doxycycline) could be considered to prevent secondary bacterial infections.

CONTACT OF THE SKIN OR EYES WITH TOXIC SUBSTANCES

If a toxic substance comes into contact with the skin or eyes, it is crucial to remove the poison as soon as possible. It is also important to prevent a bird from ingesting the toxic substance through the beak while preening.

ADVICE FOR FIRST AID AT HOME

The priority is to remove the toxic substance from the skin or eyes as soon as possible. Owners should put on gloves to prevent contact of their own skin with the toxic substance and avoid transferring the toxin to any other areas of skin, mucous membranes or eyes. Skin should be rinsed thoroughly with lukewarm tap water. Eyes should be rinsed thoroughly preferably with physiological saline solution, but if this is not available lukewarm tap water will suffice.

Excessive rinsing or flushing can lead to a dangerous decrease in body temperature (hypothermia). Therefore, birds should not be rinsed too long at home. Birds that are depressed or lethargic might already be hypothermic, so rinsing should be gentle.

Depressed or lethargic birds with toxic substances on the feathers only should not be bathed. Owners should prevent these birds from preening and ingesting the toxins by distracting or holding them until the emergency consultation.

* * *

EMERGENCY CARE BY A VETERINARY PROFESSIONAL

A neck collar can be applied to prevent birds from preening and ingesting toxic substances if holding is not possible all the time.

Poison from the eyes should be removed immediately by washing with lukewarm saline or water.

If necessary, information about the specific poison should be obtained by calling a poisoning information center.

General stabilization

Sick patients are stabilized using heat, nutrition, fluid therapy, oxygen (see General stabilization of sick birds, p. 23).

When possible, a specific treatment against the intoxication is started.

After stabilization, as much poison as possible should be removed from the skin and feathers by washing with lukewarm water and, if necessary, diluted liquid dishwashing soap (e.g., Dawn). Birds should be monitored carefully during washing, especially for signs of hypothermia. Washing should be stopped when birds become unstable again and must only be continued after fully stabilizing the patient.

In a case of contact with the eyes, the eyes should be checked for corneal damage using fluorescein. In cases of corneal defects, treatment with an eye ointment with vitamin A and antibiotics (without corticosteroids) is indicated after removal of the toxin.

LEAD POISONING

Lead poisoning is relatively common in birds and leads to life-threatening situations. The sooner treatment is started, the better the prognosis.

Birds can encounter lead and ingest lead in many households and outdoors as well. Psittacines can ingest lead by biting on lead-containing materials, such as lead curtain weights (**Fig. 22.1**), batteries, lead lights/leaded windows, lead foil on old wine bottles, and paint. Waterfowl can eat lead fishing gear (**Fig. 22.2**). Chickens can swallow almost everything while foraging, including lead particles. Birds of prey can ingest lead by eating prey that has been shot with lead shot.

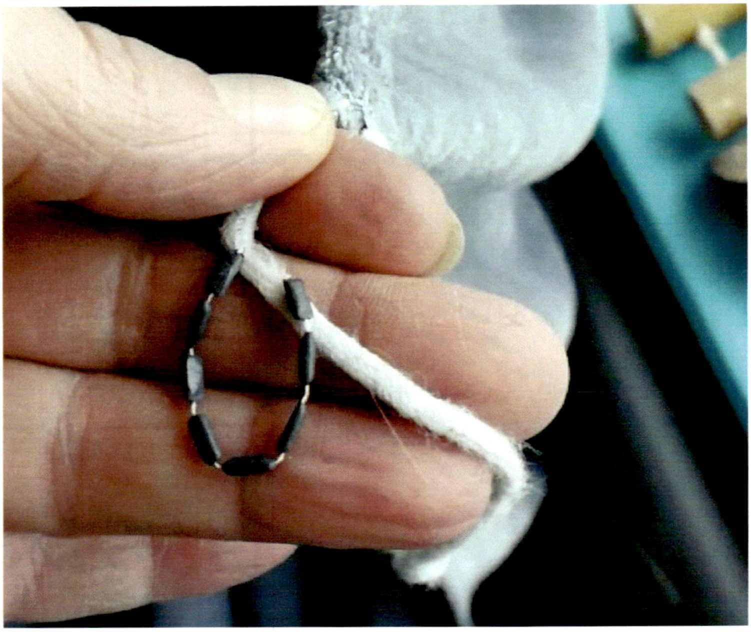

Fig. 22.1 Lead curtain weight.

It is not uncommon for owners to be unaware of lead being present in their bird's environment.

Fig. 22.2 Juvenile swan with fishing gear.

Symptoms of lead poisoning depend on the bird species and the amount of ingested lead, among other things. Symptoms can include paralysis of the toes and legs, regurgitating/vomiting, impaired coordination, lethargy, abnormal behavior, seizures, bloody droppings, slow heartbeat, blindness, and pink discoloration of the normally white part of the dropping (urates). Not every symptom will be present in every affected bird. For example, in lovebirds, paralysis of the toes can be the only symptom in early stages of lead intoxication, while in cockatiels, nausea or epilepsy can be the main symptom.

If no timely action is taken in cases of severe lead poisoning, birds will die.

* * *

EMERGENCY CARE BY A VETERINARY PROFESSIONAL

X-rays

In cases of suspected lead intoxication due to information from the medical history or clinical symptoms, X-rays (see Appendix 3, p. 179) are indicated. On X-rays, metal particles are visible as bright white fragments in the gastrointestinal tract (**Fig. 22.3**). In situations of acute lead poisoning, clear radiodense particles are usually visible on X-rays, but the absence of these particles does not rule out metal poisoning entirely. On the other hand, it is important to realize that not every metal is toxic and that the presence of metal particles on X-rays is not definitive evidence of lead poisoning.

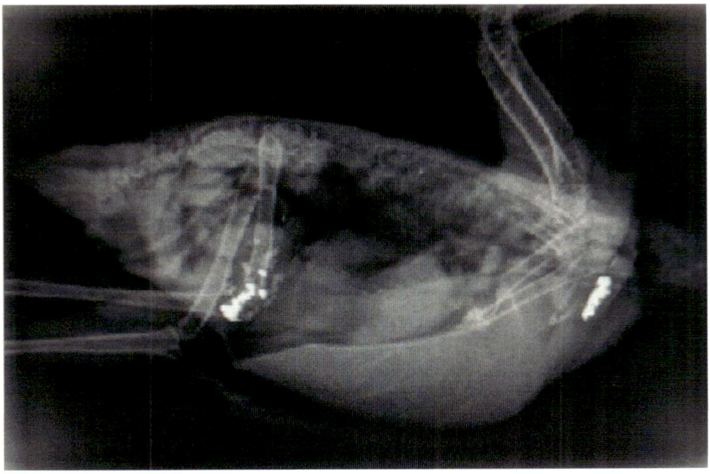

Fig. 22.3 X-ray showing multiple (white) metal particles in the crop and stomach.

Blood test

In cases of suspected lead intoxication, the lead level in the blood can be measured. See Observation, physical examination and diagnostic tests, p. 13 and Appendix 2, p. 169.

Unfortunately some laboratories require a relatively large volume of blood for determination of the lead concentration. This can exceed the volume of blood that can safely be taken from small patients.

Note: In very small species or seriously ill birds, collecting blood is sometimes too dangerous, especially by veterinarians less trained in collecting blood from birds. Too much blood loss, formation of hematomas, handling for too long or excessive stress can lead to death in small or unstable patients. In some cases, blood collection is best not done during the emergency consultation.

Treatment

In case of proven lead intoxication or a strong suspicion of lead intoxication based on the medical history, symptoms, and/or X-ray findings, treatment must be started immediately. Also, when blood has been taken for the determination of the lead concentration, treatment should be initiated while awaiting results.

General stabilization and fluid therapy

Birds with lead poisoning must be stabilized using fluid therapy, heat and nutrition (see General stabilization of sick birds, p. 23).

Due to the nephrotoxicity of lead, fluid therapy (see p. 27) is always indicated in case of lead poisoning (also in well-hydrated birds). Initially, this is done by injection, as birds with lead poisoning can suffer from hypomotility/ileus. If the gastrointestinal tract is found to be functional and the patient does not vomit and has no seizures, some of the fluid may be given orally after stabilization.

Removal of lead particles from the gastrointestinal tract

Metal particles in the crop can be removed by crop lavage (see Appendix 5, p. 195) or endoscopy in stable patients.

In case of lead particles present in the (pro)ventriculus or intestines, laxatives (e.g., lactulose) and prokinetic drugs (e.g., metoclopramide) are administered to accelerate passage through the gastrointestinal tract and reduce systemic uptake.

Chelation

Chelating agents are administered to remove absorbed lead from the body. In the acute stage, CaNaEDTA is administered for 3–5 days. Because lead can be stored in bone tissue, long-term therapy is often necessary to prevent a relapse, even when symptoms have completely resolved during treatment in the acute stage. To reduce the risk of side effects, pain on injections sites and to enable treatment of patients at home, oral penicillamine treatment can be initiated after the initial 3–5 day treatment with CaNaEDTA. Other treatment options are repeated series of treatments with CaNaEDTA (3 days on, 3 days off).

Symptomatic treatment of seizures

Midazolam can be used to suppress seizures in the acute stage.

Symptomatic treatment of vomiting/motility stimulation

Lead poisoning can cause nausea and gastrointestinal hypomotility. Metoclopramide (parenteral) is indicated to suppress nausea and to stimulate motility.

INTOXICATION BY POISONOUS PLANTS

Many birds will bite plants if the opportunity arises. Sometimes plant parts are actually ingested, in other cases, the plant parts are just chewed on. There are many types of poison in plants. In most cases, biting or swallowing poisonous plant material primarily causes gastrointestinal disorders like vomiting, anorexia or pain caused by damage to mucous membranes. However, there are also plant toxins that can cause serious liver, kidney, nerve or heart problems. Ingesting poisonous plants can lead to discomfort and even death. Birds in captivity generally have insufficient insight into which plant species are poisonous and which are not. Because of that, birds must be protected by their owners from being exposed to toxic plant species (**Fig. 22.4**).

Unfortunately, no extensive scientific research has been done on the toxicity of plants in birds, so there is little reliable data. Certain plant species have shown in practice that they are certainly harmful to birds, while other plant species are suspected to be poisonous to birds because of known toxicity to humans or other animal species, such as dogs, cats, and livestock. However, the latter proves nothing, because differences in interspecific sensitivity do exist. There are several examples of certain berries that are very poisonous to mammals that can be eaten by birds without causing disease. On the other hand, some plant species seem to be much more harmful to birds than to other species.

Fig. 22.4 Example of a pretty but poisonous plant (Arum).

See Appendix 10 (p. 217) for a list of plants known or suspected of being poisonous to birds. The sensitivity of birds to some of the plants on this list is probably overestimated and symptoms after ingestion of parts of the plant could be mild or even absent. On the other hand, the list is undoubtedly not complete and not every plant toxic to birds is included.

ADVICE FOR FIRST AID AT HOME

Although there is no standard treatment for every kind of plant intoxication, removal of visible plant parts from the beak and if possible from the mouth is always beneficial. Administration of tap water into the front of the beak with a syringe can help to dilute toxins. Owners should be very careful that the bird doesn't choke. Owners should report the name of the plant by phone. This will give the veterinarian the opportunity to collect information about the specific intoxication and start appropriate treatment as soon as the bird arrives for the emergency consultation. Pictures should be taken when the plant species is not known by the owner, so the plant can be identified later.

* * *

EMERGENCY CARE BY A VETERINARY PROFESSIONAL

The treatment depends on the specific type of poison that has been ingested. If necessary, information about the specific poison should be obtained by calling a poisoning information center.

General stabilization

Sick patients are stabilized using heat, nutrition, fluid therapy, oxygen (see General stabilization of sick birds, p. 23), and specific treatment against the intoxication, when possible.

Removal of plant parts and preventing systemic uptake

Plant parts present in the oral cavity should be removed manually. Plant parts present in the crop can sometimes be removed or at least diluted by performing a crop lavage (see Appendix 5, p. 195). Laxatives (e.g., lactulose) can help to accelerate passage through the gastrointestinal tract and reduce systemic uptake of toxins.

Symptomatic and additional treatment

Metoclopramide and/or maropitant are indicated in case of nausea. Sucralfate (administered into the beak, not through a feeding tube) is indicated after ingestion of toxins that cause irritation or ulceration of mucous membranes. Analgesics are indicated in case of irritation and ulceration of the mucous membranes of the gastrointestinal tract. In case of ulceration, NSAIDs should not be used.

Note: Administration of activated charcoal can be beneficial in some specific cases but may be contraindicated in other cases. Do not do this by default, but only on specific indication.

INGESTION OF CORROSIVE TOXINS

Corrosive chemicals such as cleaning agents ammonia and bleach cause severe burns of the beak, esophagus and crop after ingestion.

ADVICE FOR FIRST AID AT HOME

Dilution of the poison can help to minimize the effects of the intoxication. Administration of tap water (up to 2 ml/100 grams body weight) into the front of the beak with a syringe can help to dilute toxins. Owners should be very careful that the bird doesn't choke.

Owners should report the name of the ingested toxin by phone. This will give the veterinarian the opportunity to collect information about the specific intoxication and start appropriate treatment as soon as the bird arrives for the emergency consultation. When possible, owners should take the packaging of the toxin to the veterinary clinic.

* * *

EMERGENCY CARE BY A VETERINARY PROFESSIONAL

The treatment depends on the specific type of poison that has been ingested. If necessary, information about the specific poison should be obtained by calling a poisoning information center.

Note: Emesis should not be induced after ingestion of corrosive substances, as vomiting and regurgitation can lead to further damage to the esophagus when the corrosive toxin passes for the second time and to aspiration. Activated charcoal is also contraindicated after ingestion of caustic toxins.

When possible, specific treatment against the intoxication should be started.

Suppression of nausea

For the reasons mentioned previously, birds that are regurgitating/vomiting/gagging must be symptomatically treated with metoclopramide (IM) and/or maropitant (IM).

Dilution of toxins

To minimize damage caused by corrosive toxins, dilution of the toxins can be very important.

For diluting toxins in the oral cavity and cervical esophagus, water needs to be administered directly into the beak. Usually, 2 ml can be administered to a bird with a body weight of 100 grams. Care should be taken not to cause aspiration of the administered water or regurgitation. Safe administration of oral fluids can be impossible in birds that are not tame or very stressed.

For diluting toxins in the crop, water can be administered through a feeding tube (see Appendix 5, p. 195). The crop tube should be inserted with great care and after applying some lubricant (a very thin layer to prevent aspiration), as painful mucosal lesions might already be present in the throat or esophagus.

General stabilization

Sick patients are stabilized using heat, nutrition, fluid therapy and oxygen (see General stabilization of sick birds, p. 23).

Symptomatic and additional treatment

Sucralfate (administered into the beak, not through a feeding tube) is indicated after ingestion of toxins that cause irritation or ulceration of mucous membranes. Analgesics are indicated. In case of ulceration, NSAIDs should not be used.

Most birds will need supportive care, including fluid therapy, nutritional support, antiemetics and analgesics for longer periods after ingestion of corrosive toxins.

INGESTION OF OTHER TOXIC SUBSTANCES

There are many more toxins that birds can ingest, like avocado, alcohol, drugs, chocolate, pesticides, rodent poison, snail poison, etc. The treatment after eating toxic material depends entirely on the type of poison.

ADVICE FOR FIRST AID AT HOME

Owners should report the name of the ingested toxin by phone. This will give the veterinarian the opportunity to collect information about the specific intoxication and start appropriate treatment as soon as the bird arrives for the emergency consultation. When possible, owners should take the packaging of the toxin to the veterinary clinic.

* * *

EMERGENCY CARE BY A VETERINARY PROFESSIONAL

The treatment depends on the specific type of poison that has been ingested. If necessary, information about the specific poison should be obtained by calling a poisoning information center.

When possible, specific treatment against the intoxication should be started based on the advice of the poisoning information center.

General stabilization

Sick patients are stabilized using heat, nutrition, fluid therapy and oxygen (see General stabilization of sick birds, p. 23).

Removal of toxins and preventing systemic uptake

Toxic substances present in the oral cavity should be removed manually.

When ingested toxic substances are still in the crop, performing a crop lavage (see Appendix 5, p. 195) can be beneficial for removing or diluting toxins. However, with specific intoxications, performing a crop lavage is contraindicated; this should not be done by default, but only on specific indication.

Note: With certain intoxications administration of activated charcoal into the crop through a feeding tube can help minimize the systemic uptake of toxins. However, this is contraindicated in cases of other specific intoxications (e.g., strong acids, strong bases and petroleum products). This should not be done by default, but only on specific indication.

The passage of toxins through the gastrointestinal tract can be accelerated by administering laxatives (e.g., lactulose). This can reduce absorption.

Symptomatic and additional treatment

Metoclopramide and/or maropitant are indicated in case of nausea. Sucralfate (administered into the beak, not through a feeding tube) is indicated after ingestion of toxins that cause irritation or ulceration of mucous membranes. Analgesics are indicated in case of irritation and ulceration of the mucous membranes of the gastrointestinal tract. In case of ulceration, NSAIDs should not be used.

23 *Concussion*

FLYING into a window or other object at full speed can lead to a concussion. A concussion is a form of traumatic brain injury. As a result, birds can be lethargic, have impaired coordination and/or have an abnormal position of the head or limbs.

ADVICE FOR FIRST AID AT HOME

Birds should be placed in a dark and quiet environment at room temperature for a short period of time (for example 15 minutes). Placing the patient under a heat lamp can lead to an exacerbation of damage in the event of brain injury!

EMERGENCY CARE BY A VETERINARY PROFESSIONAL

Birds with concussions should be stabilized using fluid therapy, nutrition and oxygen (see General stabilization of sick birds, p. 28). Heating of patients with concussions must be avoided to prevent further secondary brain injuries. Anti-inflammatory drugs may be indicated for brain swelling and headaches. Corticosteroids can cause serious side effects in birds and are therefore not recommended. Meloxicam can be given as an analgesic anti-inflammatory drug. Birds with concussions should be placed in quiet and silent places with dimmed lights.

DOI: 10.1201/9781003308270-26

24 *Cloacal prolapse*

IN THE CLOACA, the gastrointestinal tract, urinary tract and reproductive tract come together. A prolapse can involve multiple structures: the cloaca itself, the oviduct and/or the intestine.

Fig. 24.1 Prolapse (with an egg inside).

Regardless of the cause, complications rapidly develop in cases of prolapse due to swelling, damage, self-mutilation and infection of prolapsed tissues (**Fig. 24.1**).

ADVICE FOR FIRST AID AT HOME

Prolapsed tissues can be cleaned by flushing it with physiological saline solution. Owners should not rub with a towel, as prolapsed tissues can be very fragile. The bird should then be kept on a clean damp cloth until the prolapse has been treated by a veterinarian. Repeated moistening of the prolapse with saline or lubricant can prevent dehydration of the prolapsed tissues and thereby reduce the risk of complications. Biting or pecking the prolapsed tissues by patient or other birds (especially chickens) should be prevented by isolating the bird and distracting or holding it.

* * *

DOI: 10.1201/9781003308270-27

EMERGENCY CARE BY A VETERINARY PROFESSIONAL

Abdominal palpation, X-rays (see Appendix 3, p. 179) and gentle cloacal exploration/palpation with a finger in large birds and cotton tip applicator in small birds should be attempted to determine the cause of the prolapse. In cases of space-occupying masses, such as cloacal stones (cloacoliths) or eggs (**Fig. 24.1**), removing these is necessary before repositioning the displaced structures and resolving the prolapse.

GENERAL STABILIZATION

Birds with prolapses are generally weakened and must be stabilized using heat, fluid therapy, food and, in case of dyspnea, oxygen (see General stabilization of sick birds, p. 23).

ANALGESIA

Prolapses are very painful and analgesia (e.g., meloxicam and/or butorphanol) is indicated. Adequate analgesia can also result in a decrease of tenesmus, making managing the prolapse more rewarding.

MANAGEMENT OF THE PROLAPSE

Besides maybe the initial cleaning with physiological saline solution, treatment is done under anesthesia (e.g., with isoflurane/sevoflurane). First, the prolapsed tissues must be cleaned gently but thoroughly. In case of severe swelling of the prolapsed tissues, an attempt can be made to reduce the swelling by means of hypertonic glucose solution. Second, the prolapsed tissues should be returned to their normal position using gloved fingers or cotton tipped applicators. Water-based lubricant should be used and care should be taken not to cause additional damage to the fragile tissues. To counteract immediate recurrence, the size of the opening of the cloaca can be reduced by means of horizontal mattress sutures on both sides of the cloacal opening (**Fig. 24.2**). A central opening should remain, leaving just enough space for passage of droppings.

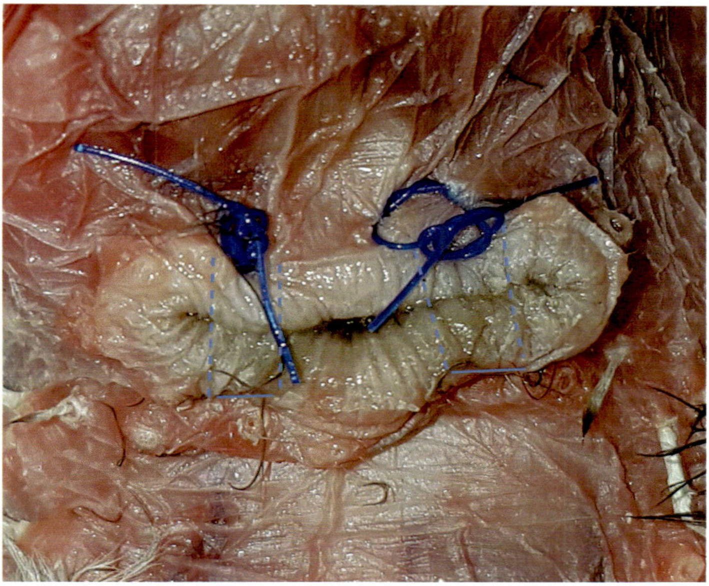

Fig. 24.2 Sutures placed to reduce the size of the cloacal opening. The blue lines indicate the suture pattern.

The cloaca and sutures should be checked regularly for recurrence of the prolapse and possible complications (for example dystocia, constipation and infection).

ANTIBIOTICS

Systemic broad-spectrum antibiotics (e.g., amoxicillin/clavulanic acid) are indicated in case of traumatized, devitalized and/or infected tissues.

Unless the cause of the prolapse has been diagnosed and the prolapse and underlying cause have been adequately treated, further diagnostics and treatment is required.

25 *Vomiting*

VOMITING can be a symptom of many conditions, some relatively harmless and others life threatening. Vomiting can quickly lead to severe dehydration and/or starvation. The cause of vomiting can be a disorder inside or outside the gastrointestinal tract. Causes within the gastrointestinal tract include infections, foreign objects, motility disorders, tumors or intussusception. Examples of non-gastrointestinal causes of vomiting include intoxications, liver or kidney diseases and disorders of the central nervous system.

Note: When a bird actively gives up ingested food (regurgitation) aimed at the partner, owner, toy, mirror or other object, this is probably "partner behavior". In this case, there is no vomiting or threatening emergency situation. However, repeated regurgitation in the absence of a 'recipient' may indicate underlying infection or other issue.

EMERGENCY CARE BY A VETERINARY PROFESSIONAL

Severe causes of vomiting are not uncommon. For example, lead poisoning is a frequently occurring cause of vomiting, especially in psittacines, backyard poultry and waterfowl. Because of the possibility of acutely life-threatening diseases, performing diagnostics is always recommended in vomiting birds instead of just symptomatic treatment. Suppressing vomiting without identifying and solving the underlying disorder might lead to a worsening of the situation and even death.

ANAMNESIS

Information should be obtained from the owner about the possibility of ingestion of poisonous materials (e.g., toxic plants, drugs or lead). Most owners are unaware of the presence of lead in the environment of birds, so lead poisoning cannot be ruled out based on anamnesis alone. For more information about intoxications, see Intoxications, p. 71.

CROP PALPATION

The crop is gently palpated and checked for filling and the presence of foreign bodies. A filled crop in nauseous birds can be an indication of crop stasis/sour crop (see Crop stasis, p. 95). Foreign bodies (*corpora aliena*) can be single solid objects, but also include substrate, grass (notorious in chickens), hair, rope or textiles.

FECAL EXAMINATION

Fecal samples are taken for microscopic examination (see Appendix 4).

If a complete fecal examination during the emergency consultation is not possible, fecal samples and smears are kept for later examination.

DOI: 10.1201/9781003308270-28

Note: Birds of most species do have bacterial intestinal flora, so the presence of bacteria in the feces of most birds is normal.

MICROSCOPIC EXAMINATION OF CROP CONTENTS

Microscopic examination is performed on crop contents or fresh vomit. Fresh samples are diluted with warm saline and checked for the presence of moving flagellates. Thin smears on microscopic slides are stained and checked for large numbers of bacteria or yeasts. Crop contents are not sterile in healthy birds, so presence of some bacteria and yeasts is not abnormal (**Fig. 25.1**). Very large amounts of bacteria (especially monocultures, **Fig 25.2**) and/or yeasts (especially budding yeasts or yeasts forming pseudohyphae, **Fig. 25.3**) are indicative of overgrowth of these microorganisms.

Note: Primary crop infection is not very common and overgrowth of microorganisms is often secondary to an underlying (motility) disorder. In cases of secondary overgrowth of microorganisms, treatment for this alone is not sufficient. The underlying cause should be determined and solved. If this is not possible during the emergency consultation, birds can be referred to an avian specialist after stabilization for further diagnostics and treatment.

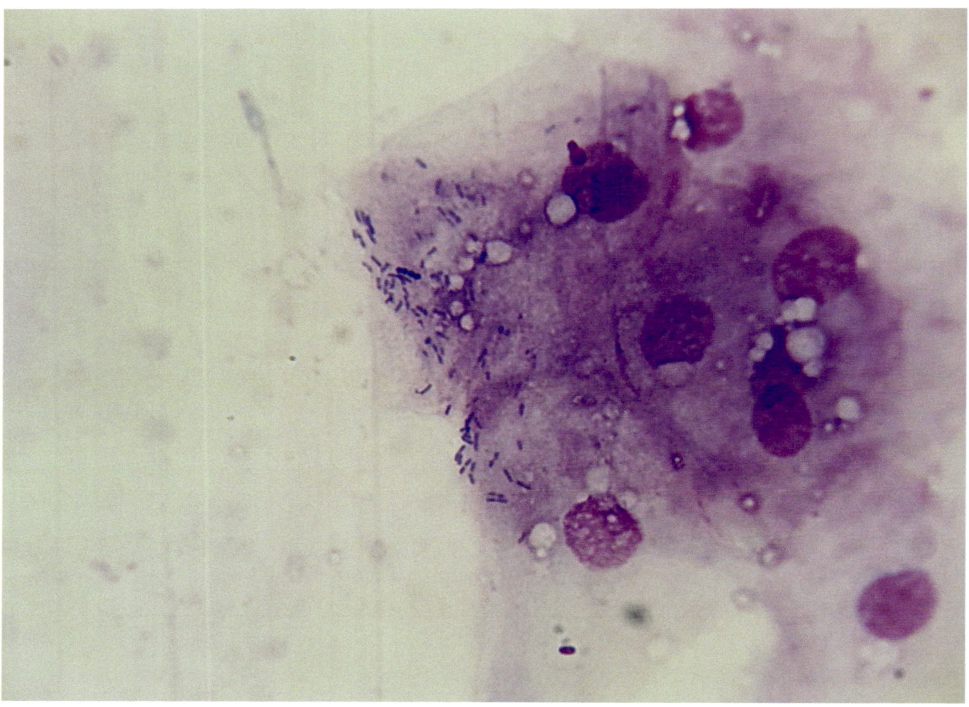

Fig. 25.1 Normal cytology of the crop: Epithelial cells, mixed bacterial culture and a single yeast.

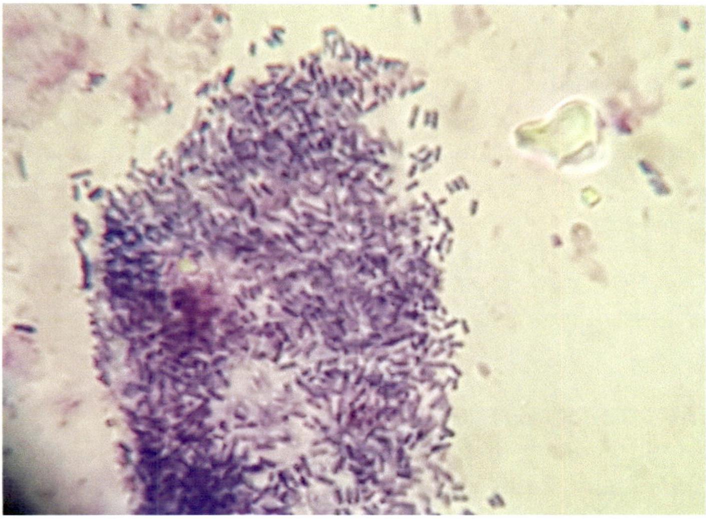

Fig. 25.2 Abnormal cytology of the crop: Bacterial overgrowth, monoculture.

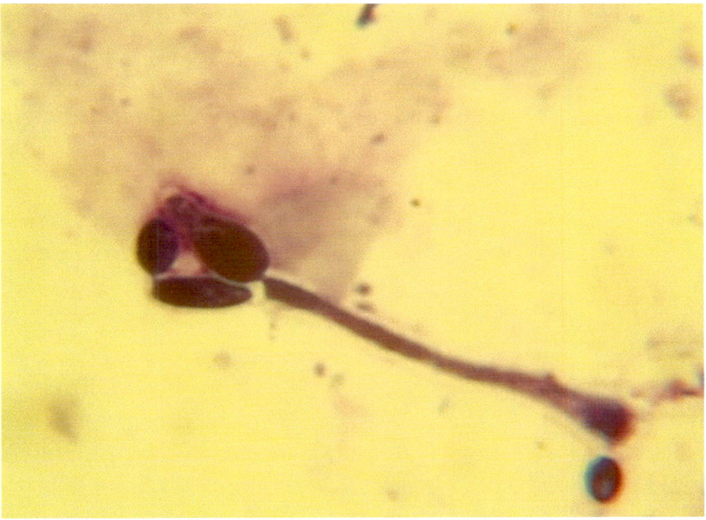

Fig. 25.3 Abnormal cytology of the crop: Yeast forming pseudohyphae.

X-RAYS

X-rays (see Appendix 3, p. 179) are indicated to check for the presence of radiodense foreign bodies, space-occupying lesions or metal particles (see Lead poisoning, p. 73).

Note: Lead poisoning cannot be completely ruled out even if metal particles are not visible on X-rays, and if there is a strong suspicion of such, based on anamnesis or clinical symptoms, then treatment must be started quickly, pending the results of blood tests.

BLOOD TESTS

Vomiting can be the result of disorders outside the gastrointestinal tract and can cause complications including dehydration and changes in electrolyte concentrations. Ideally, blood should be collected (see Observation, physical examination and diagnostic tests, p. 13 and Appendix 2, p. 169) for hematology and biochemistry.

Note: In very small species or seriously ill birds, collecting blood is sometimes too dangerous, especially by veterinarians less trained in collecting blood from birds. Too much blood loss, formation of hematomas, handling for too long or excessive stress can lead to death in small or unstable patients. In some cases, blood collection is best not done during the emergency consultation.

SYMPTOMATIC TREATMENT OF VOMITING

To counteract (further) dehydration, loss of nutrients and to enable effective oral treatment, vomiting/regurgitation should be stopped. Metoclopramide is effective in most cases. Until the vomiting/regurgitation has completely stopped, administration should be IM or IV. After successfully stopping vomiting, further dosing can be performed orally, with the exception of birds suffering from crop stasis (see p. 95), in which a delayed emptying of the crop prevents systemic uptake of medication given orally.

Note: Metoclopramide is contraindicated in case of gastrointestinal bleeding and intestinal obstruction.

Maropitant can be used in addition to metoclopramide, if necessary, or instead of metoclopramide when metoclopramide is contraindicated. Maropitant does not have prokinetic properties.

GENERAL STABILIZATION

Vomiting birds must be stabilized using fluid therapy, heat and nutrition (see General stabilization of sick birds, p. 23). Fluid therapy should be parenteral as long as vomiting persists. Once vomiting/regurgitation has stopped by suppressing nausea, nutrition should be offered. Because sick birds are frequently anorectic (for reasons other than nausea), tube feeding is indicated in many cases after the vomiting/regurgitation has stopped.

TREATMENT OF MICROORGANISM OVERGROWTH

In case of bacterial overgrowth, antibiotics are indicated, e.g., amoxicillin/clavulanic acid (doxycycline and TMPS often cause nausea as a side effect, which is undesirable in already-vomiting birds).

In yeast overgrowth, antifungal agents are indicated. Nystatin (PO) and amphotericin B (PO) are generally effective in cases of yeast overgrowth in the crop and are relatively safe to use because of the minimal systemic absorption after oral intake. Itraconazole can be used as well, but it is systemically absorbed and can cause more side effects. Macrorhabdus (megabacteriosis) can be treated with amphotericin B (PO).

Flagellates can be treated with oral antiprotozoal drugs from the nitroimidazoles class (e.g., metronidazole, ronidazole). Worm infections can be treated with anthelmintics (e.g., fenbendazole, flubendazole, ivermectin, praziquantel). Coccidia can be treated with anticoccidial drugs (e.g., toltrazuril, sulphadimethoxine).

TREATMENT IN CASE OF A FOREIGN BODY

In the case of foreign bodies other than metal or sharp objects in the crop, the recommended approach is to suppress nausea (metoclopramide) and to stabilize the patient first (see General stabilization of sick birds, p. 23).

Note: When large objects are present in the crop, only smaller volumes of fluids/liquid food can be administered orally because of the risk of regurgitation.

After stabilization, foreign objects should be removed. In case of metallic foreign bodies, removal can be attempted by introducing a magnet glued to, for example, a feeding tube into the crop (**Figs. 25.4 and 25.5**). Unfortunately, not every metal is magnetic

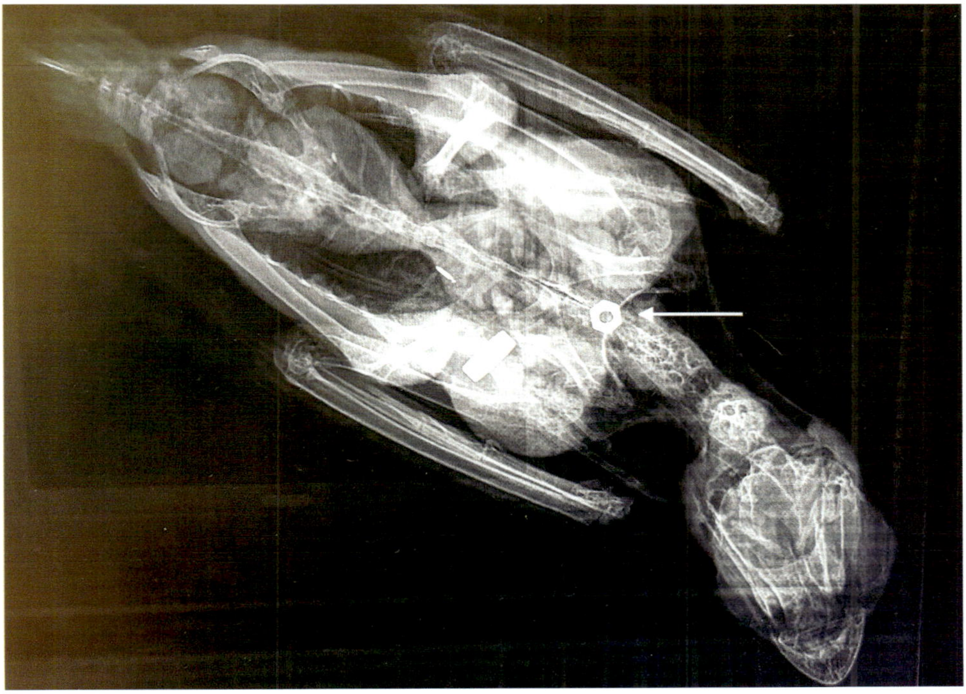

Fig. 25.4 Metal nut (arrow) in crop of an eclectus parrot.

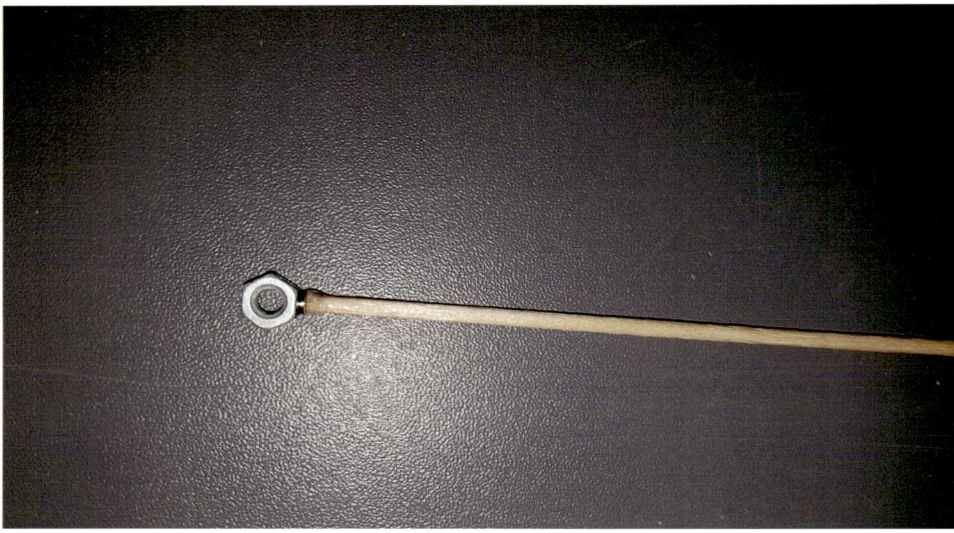

Fig. 25.5 Nut removed by introducing a strong magnet glued to a stick into the crop (under general anesthesia).

Sharp, non-magnetic and/or big objects should be removed from the crop by endoscopy or surgery (see Appendix 9, p. 213).

26 *Crop stasis*

CROP STASIS is caused by obstruction or motility disorders of the crop, and is noted when the crop fails to empty and becomes distended. Because of this disorder, food remains in the crop for too long, and secondary overgrowth of bacteria or yeasts is common (sour crop). Especially in birds of prey, sour crop with rotting meat is an acutely life-threatening condition. The crop does not have a function in absorption of water or nutrients, so crop stasis (with no fluids or food moving to the stomach and intestines) results in dehydration and starvation.

Externally, a swelling at the base of the neck can be seen and/or felt (**Fig. 26.1**). The patient may be nauseated and weak.

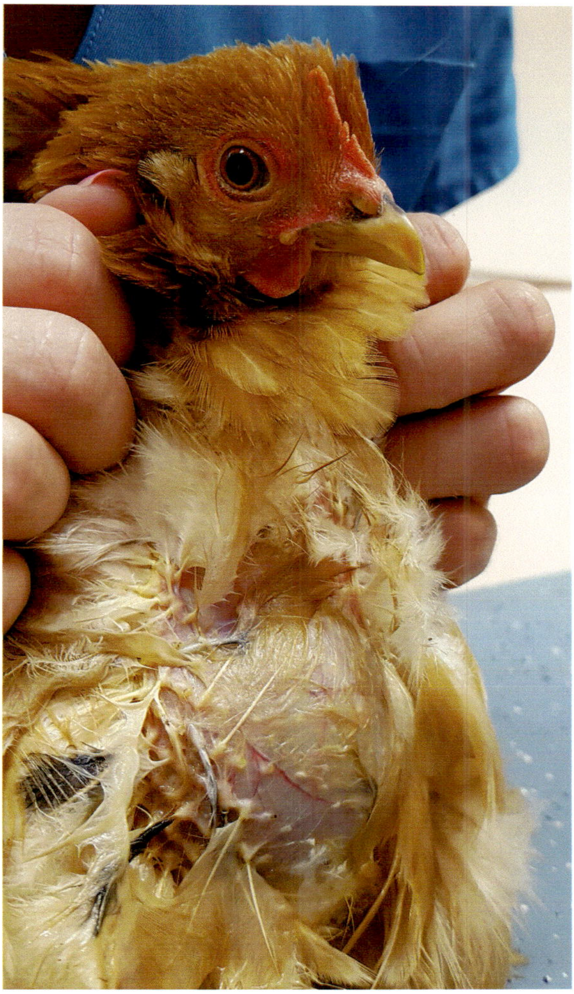

Fig. 26.1 Chicken with crop stasis. Note the swelling at the base of the neck.

DOI: 10.1201/9781003308270-29

Fig. 26.2 Grass removed from the crop of a chicken with crop impaction.

Crop stasis can be caused by disorders of the crop itself (infections of the crop or the presence of swallowed foreign material, **Fig. 26.2**), but also by disorders of other parts of the gastrointestinal tract or disorders outside the gastrointestinal tract. For example, dehydration, hypothermia and neurological disorders (such as a lead poisoning, Marek's disease or Avian Bornaviral Ganglioneuritis) can also contribute to reduced motility and emptying of the crop.

ADVICE FOR FIRST AID AT HOME

In cases of hard crop contents in bigger species like chickens, administering some lukewarm tap water orally followed by a gentle massage of the crop can help to soften the crop contents, facilitating passage to the proventriculus. This procedure is not without risks (owners should be very careful that birds don't aspirate the water and choke) and is, in most cases, not effective as a sole treatment. Therefore, this procedure is not advised.

Note: When the crop is heavily filled with liquid content, massaging the crop is contraindicated even without the administration of extra water! It can cause regurgitating and aspiration leading to severe complications or even death.

Birds with crop stasis should be kept warm until the bird is at the veterinary clinic for examination and treatment.

* * *

EMERGENCY CARE BY A VETERINARY PROFESSIONAL

CROP PALPATION

The crop is gently (to prevent regurgitation) palpated and checked for solid or fluid contents or foreign bodies.

MICROSCOPIC EXAMINATION

Microscopic examination is performed on crop contents or fresh vomit (see Vomiting, p. 89)

Note: Primary crop infection is not very common and overgrowth of microorganisms is often secondary to underlying (motility) disorders. In the case of secondary overgrowth of microorganisms, treatment for this alone is not sufficient. The underlying cause should be determined and solved. If this is not possible during the emergency consultation, the bird should be referred to an avian veterinarian after stabilization.

X-RAYS

X-rays (see Appendix 3, p. 179) are indicated to check for the presence of radiodense foreign bodies, space occupying lesions or metal particles (see Lead poisoning, p. 73).

In the case of liquid crop contents, X-ray examination in a perfectly positioned (sedated) lying down bird can be dangerous because of the risk of regurgitation and aspiration (this risk can be minimized by intubation of the trachea). In this situation, taking vertical and preferably horizontal radiographs of a conscious, standing bird (in a radiolucent container) is a safer option. While this imperfect positioning results in the loss of visibility of many body structures on radiographs, contents of the crop and rest of the gastrointestinal tract can still be examined.

Note: Lead poisoning cannot be completely ruled out even when no metal particles are visible on X-rays. If there is a strong suspicion of lead poisoning based on anamnesis or clinical symptoms, treatment should be started quickly, preferably pending the results of blood tests.

TREATMENT OF CROP STASIS IN NON-RAPTORS

When crop contents are very liquid, emptying the crop by suction through a crop tube (see Appendix 5, p. 195) can be successful.

TREATMENT OF INFECTION OR OVERGROWTH OF MICROORGANISMS

Flagellates can be treated with oral antiparasitic drugs from the nitroimidazoles class (for example metronidazole). In case of bacterial overgrowth in the crop, antibiotics are indicated, e.g., amoxicillin/clavulanic acid. In case of yeast overgrowth in the crop, antifungal agents are indicated. Nystatin and amphotericin B are generally effective in case of yeast overgrowth in the crop and are relatively safe to use because of the minimal systemic absorption after oral intake. Itraconazole can be used as well, but is systemically absorbed and can cause more side effects.

Note: If it is not possible to directly assess the presence of microorganisms by means of microscopic examination during the emergency consultation, starting immediate treatment with oral antibiotics in combination with antifungal agents is advisable in case of crop stasis.

In case of crop stasis, medication for treating overgrowth of microorganisms in the crop should be given orally or directly into the crop.

STIMULATION OF GASTROINTESTINAL MOTILITY

Metoclopramide is indicated to stimulate the motility of the gastrointestinal tract. In case of crop stasis, metoclopramide should be administered by IM or IV injection.

FLUID THERAPY

Dehydration can be the cause as well as a complication of crop stasis. To break this vicious cycle and to stabilize birds with crop stasis, fluid therapy is essential. While administering oral fluids can be useful in softening crop contents, it does not contribute to restoring the hydration status of the body in case of crop stasis. Fluids should therefore be administered parenterally (see Fluid therapy, p. 27).

SOFTENING THICKENED CROP CONTENTS

Fluids administered orally can help soften thickened crop contents. For this purpose, a small amount of saline can be administered into the crop every hour through a crop tube, followed by a gentle (to prevent regurgitation) massage of the crop. How much saline can be administered safely depends on the volume of the already present crop contents, in most cases 5–10 ml/kg should at least be possible. Some avian specialists use cola instead of saline for this purpose.

Note: As noted above: When the crop is heavily filled with liquid content, massaging the crop is contraindicated! It can cause regurgitating and aspiration leading to severe complications or even death.

HEAT

Like dehydration, hypothermia can be the cause as well as a complication of crop stasis. To break this vicious cycle and to stabilize birds with crop stasis, a warm environment is essential. Birds with crop stasis must be properly supported by an external heat source (see Heat, p. 25).

NUTRITION

Feeding birds with crop stasis is not useful. Nutrients in the crop will only stimulate growth of harmful microorganisms. Start by feeding liquid food through a feeding tube (see Nutritional support, p. 33) when crop motility improves and liquids pass to the stomach. Test the passage from the crop to the stomach first with water or saline when crop motility returns, either by oral administration with a syringe or via a crop tube (see Appendix 5, p. 195).

FOREIGN BODY

Foreign bodies should be removed by endoscopy or surgery (Appendix 9, p. 213) after stabilization of the patient.

CROP STASIS IN BIRDS OF PREY

In birds of prey, sour crop is a very dangerous condition due to the decomposition of meat in the crop. Crop contents must be removed as soon as possible. This can be done by flushing the crop of an anesthetized bird. As birds of prey with crop stasis can be severely weakened, this procedure is not without risks. After isoflurane/sevoflurane induction with a mask, an endotracheal tube is placed in the trachea to prevent aspiration. Second, a flexible or rigid feeding tube is inserted through the beak into the crop (see Appendix 5, p. 195). Because birds of prey swallow larger pieces of meat, it is not possible to remove the crop contents through the feeding tube. Instead, saline or tap water at body temperature is administered into the crop and crop contents are massaged up into the cervical esophagus and then out of the beak. If this is not possible, the crop can also be emptied by endoscopy or ingluviotomy (see Appendix 9, p. 213).

Note: Birds of prey with sour crop can be seriously ill or even in shock. In case of shock, subcutaneous fluid therapy is not useful due to poor absorption. Instead, intravenous or intraosseous fluid therapy is indicated.

NUTRITION

Feeding birds of prey with crop stasis makes the situation worse. Meat in the crop will only stimulate growth of harmful microorganisms. Feeding birds of prey should be avoided until motility is restored. The emptying of the crop and passage of fluids to the stomach is tested by administering saline into the crop through a feeding tube (see Appendix 5, p. 195) and palpating crop contents regularly. Only when the passage of electrolyte solutions is normal should feeding be started. Due to the risk of recurrence of crop stasis, the consistency and volume of administered food is only gradually increased. The first meal should be a liquid meat paste. When this passes to the stomach without complications, the consistency of the food can be made a bit more concentrated the next time. The consistency of the food is gradually increased until normal parts of meat can be given.

27 *Seizures*

EPILEPTIC seizures can have many different causes. A distinction is made between primary and secondary epilepsy. Primary epilepsy is a congenital disorder with no apparent underling cause. Secondary epilepsy occurs as a result of an underlying diagnosed condition. As in mammals, both primary and secondary epilepsy occur in birds, though secondary epilepsy is more common. Frequently occurring causes of secondary epilepsy are intoxications (e.g., lead poisoning, see p. 73), cardiovascular diseases, infections of the brain, hypocalcemia (see Appendix 12, p. 227), trauma, glucose deficiency and severe liver or kidney diseases.

ADVICE FOR FIRST AID AT HOME

Birds with seizures should be placed in a small cage or box with a soft substrate (e.g., towels) to prevent falling from height and damaging the wing tips during the attacks. Sand is not a good substrate, because it might get into the eyes during attacks.

In small birds developing epilepsy after not eating for a long period of time (in this situation the droppings will contain no feces or tarry black feces), the cause may be hypoglycemia. In this situation, 15% dextrose solution can be administered into the beak when seizures have stopped for a while.

Note: When birds still have seizures or are not fully awake, administering oral medication, food or fluids into the beak is contraindicated, as it can lead to aspiration and death. In those cases, absolutely nothing may be administered into the beak!

* * *

EMERGENCY CARE BY A VETERINARY PROFESSIONAL

During the emergency consultation, attempts should be made to stabilize the patient and to suppress further seizures. In addition, any acutely threatening causes of secondary epilepsy must be diagnosed and treated. Owners are asked about possible trauma, diet, living conditions and exposure to toxic substances (e.g., lead, medication or drugs).

X-RAYS

A relatively common cause of epileptic seizures is lead poisoning (see Lead poisoning, p. 73). X-rays (see Appendix 3, p. 179) are indicated for birds with seizures to check for the presence of metal particles in the gastrointestinal tract.

DOI: 10.1201/9781003308270-30

BLOOD TESTS

Ideally, blood should be collected (see Observation, physical examination and diagnostic tests, p. 13 and Appendix 2, p. 169) for hematology, biochemistry and toxicology (e.g., lead concentration).

Note: In very small species or seriously ill birds, collecting blood is sometimes too dangerous, especially by veterinarians less trained in collecting blood from birds. Too much blood loss, formation of hematomas, handling for too long or excessive stress can lead to death in small or unstable patients. In some cases, blood collection is best not done during the emergency consultation.

GENERAL STABILIZATION

Birds with seizures must be stabilized using fluid therapy, heat, nutrition and/or oxygen (see General stabilization of sick birds, p. 23).

Seizures can lead to regurgitation and aspiration of the crop contents and death. As long as the seizures are not effectively suppressed, oral fluid therapy is too dangerous and therefore contraindicated. Parenteral fluid therapy is indicated at this stage.

Seizures can occur due to hypoxia and/or circulation problems. If hypoxia/ cardiovascular diseases have not been ruled out, then birds with epilepsy should be placed in an oxygen cage.

Birds with seizures can be hypothermic and in need of extra heat. On the other hand, birds with ongoing spasms might develop hyperthermia (see Hyperthermia, p. 49), in which case additional heat is contraindicated. Birds suspected of brain trauma should also not be overheated. This can increase secondary brain damage. Uncontrolled cooling also involves too many risks in birds. Room temperature is probably safest in these cases.

Birds that do not eat should be force-fed, unless they are vomiting (see p. 89) or the seizures have not yet been effectively suppressed, as epileptic seizures can lead to regurgitation with aspiration of the crop contents and death.

In birds suspected of hypoglycemia (in which case the droppings will usually contain either no feces or tarry black feces) 5% glucose/NaCl solution can be administered as a bolus (see Fluid therapy, p. 27). When seizures have stopped for a while, a small amount (for example 1 ml per 100 grams of body weight) of 5% glucose solution can be administered into the beak or crop through a feeding tube (see Appendix 5, p. 195).

SYMPTOMATIC TREATMENT OF SEIZURES

In case of status epilepticus, frequent and/or prolonged seizures, midazolam can be administered IV or IM to suppress the seizures.

Note: For long-term suppression of seizures, other antiepileptic agents (e.g., phenobarbital, levetiracetam) are indicated.

HYPOCALCEMIA

In cases of suspected hypocalcemia, treatment is best started immediately. Also, when blood has been taken for the determination of the (ionized) calcium concentration, treatment should be initiated while awaiting results, as hypocalcemia can lead to rapid deterioration and death.

Calcium (boro)gluconate and vitamin D3 can be administered by intramuscular injection. After stabilization, oral calcium supplements can be administered. See Appendix 12.

Note: In case of secondary epilepsy, stabilizing patients and suppressing seizures is not enough, as underlying diseases can become more severe and cause more problems in the future. In many cases, additional diagnostic tests (e.g., CT scan, hematology, blood biochemistry, ECG and cardiac ultrasound) are necessary to diagnose or exclude possible underlying diseases.

28 *Egg binding/dystocia*

EGG BINDING, or dystocia, is the disorder in which a bird is unable to expel an egg from her reproductive tract. Egg binding can be caused by abnormal eggs, disorders of the reproductive tract or disorders outside the reproductive tract. Eggs can have abnormal sizes, shapes or shell quality. Disorders of the reproductive tract include inflammation, infection, torsion, neoplasia and strictures. Disorders outside the reproductive tract include disorders of calcium metabolism (see Appendix 12, p. 227), abdominal hernia, obesity, inappropriate housing, inappropriate diet and space-occupying lesions.

Signs of egg binding include a swollen coelom/abdomen, shortness of breath (**Fig. 28.1**), bulging vent, tenesmus (ineffective straining), fresh blood in the stool or on the vent, sitting on the floor, paralysis of legs, constipation, walking strangely and sitting wide-legged. General signs such as lethargy, sitting still with fluffed feathers and a decreased appetite may also be noted. When untreated, egg binding can lead to rapid deterioration and death.

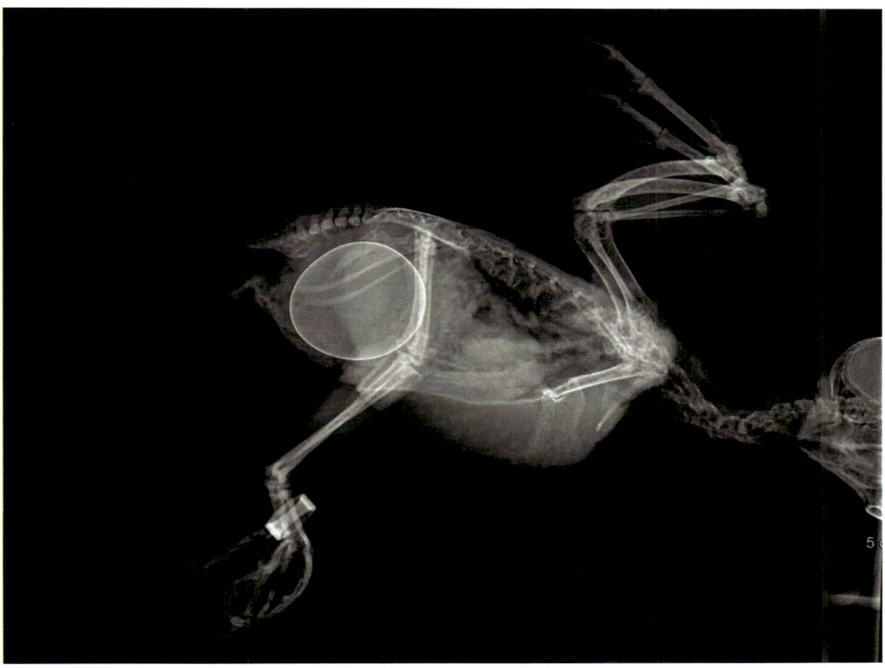

Fig. 28.1 X-ray of a bird with egg binding. The egg causes compression of the caudal air sacs and, thereby, dyspnea.

DOI: 10.1201/9781003308270-31

ADVICE FOR FIRST AID AT HOME

Egg-bound birds should be placed in a warm, quiet environment. When an egg is visible in the cloaca, the mucous membranes stretched over the egg can be moistened with physiological saline or lubricant. A very small amount of sunflower oil/olive oil can also be applied to the cloaca to function as a lubricant. Even when eggs are passed after this initial treatment, veterinary help is necessary. Risks of recurrence of egg binding with the next egg are great if no treatment is started or adjustments in diet and husbandry are made.

EMERGENCY CARE BY A VETERINARY PROFESSIONAL

Birds with egg binding are often short of breath, dehydrated and severely weakened. Paresis or paralysis of legs can occur due to pressure by the egg on nerves in the pelvic area. Handling should therefore be minimized. Place severely dyspneic birds in an oxygen cage prior to examination.

During the physical examination, swelling of the coelom/abdomen can be noted. A hard, egg-shaped mass can often be felt in the caudal part of the coelom in birds with egg binding. Do not confuse the more cranially located gizzard (ventriculus) with an egg and realize that eggs without fully calcified—and therefore soft—shells cannot be differentiated from other soft tissues by palpation.

X-RAYS

X-rays (see Appendix 3, p. 179) must be taken in case of (suspected) egg binding, even when eggs can be felt or even seen. X-rays can confirm the diagnosis of egg binding and provide information about how many eggs are involved (**Fig. 28.2**) and the size, shape, shell thickness and location of the eggs.

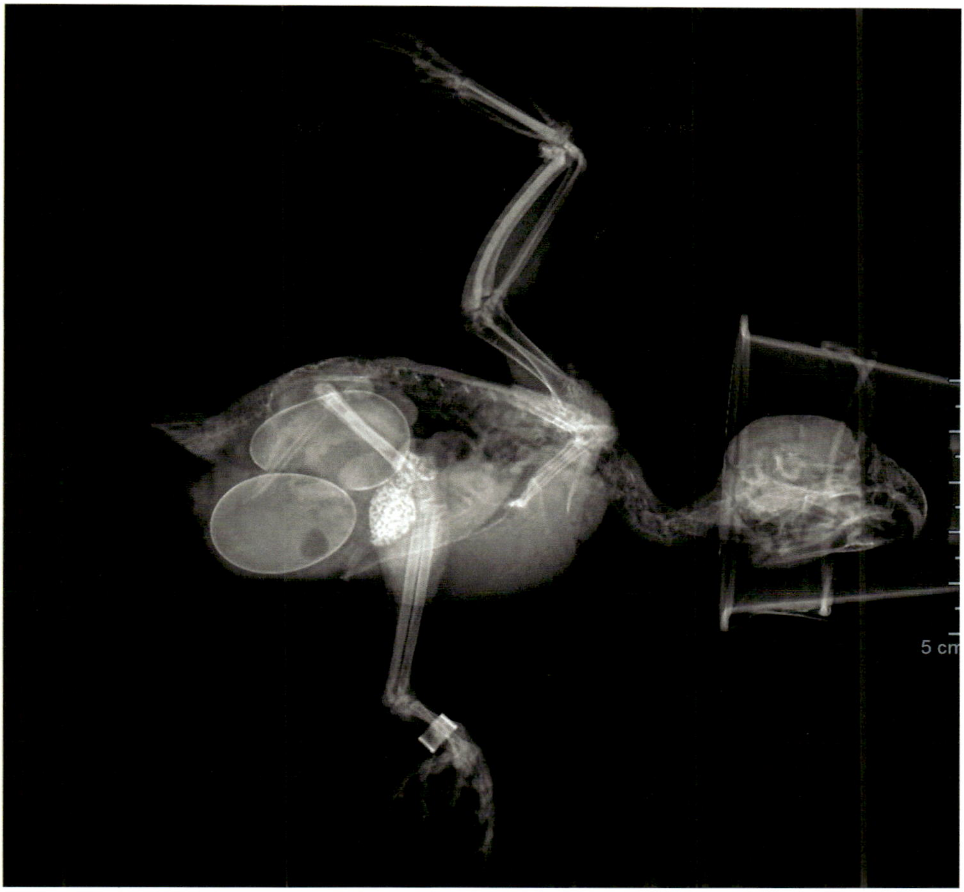

Fig. 28.2 X-ray of a quaker parrot with egg binding with two eggs, emphasizing the importance of radiology for ascertaining the exact situation.

In addition, X-rays provide information about calcium metabolism. In healthy female birds, medullary cavities of hollow bones are used for storage of extra calcium before laying eggs. This shows on X-ray as an increased radiodensity (hyperostosis, **Fig. 28.3**) of the long bones (e.g., humerus and femur). Hyperostosis occurs under the influence of female sex hormones and can be physiological or pathological.

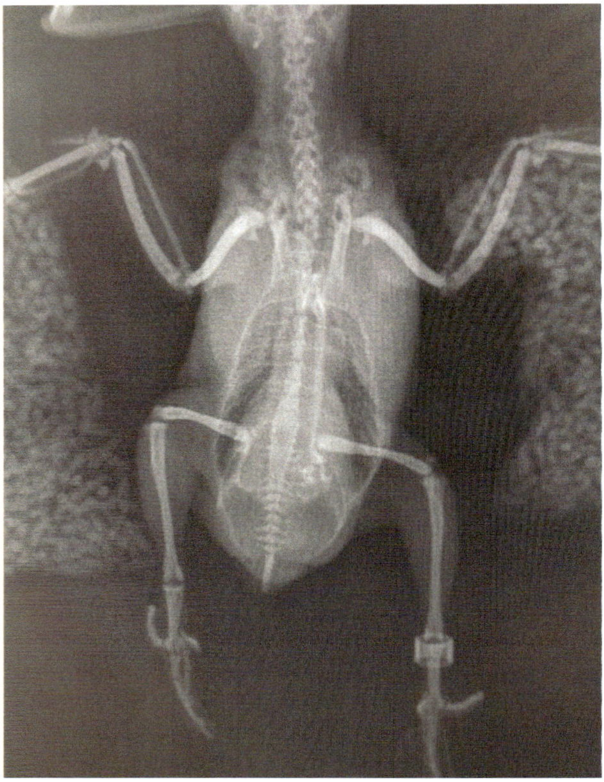

Fig. 28.3 X-ray of a bird with polyostotic hyperostosis. Note the increased medullary bone density of the long bones.

When no hyperostosis is visible on X-ray in birds with egg binding (**Fig. 28.4**), this is an indication of a disorder of calcium metabolism being the underlying cause.

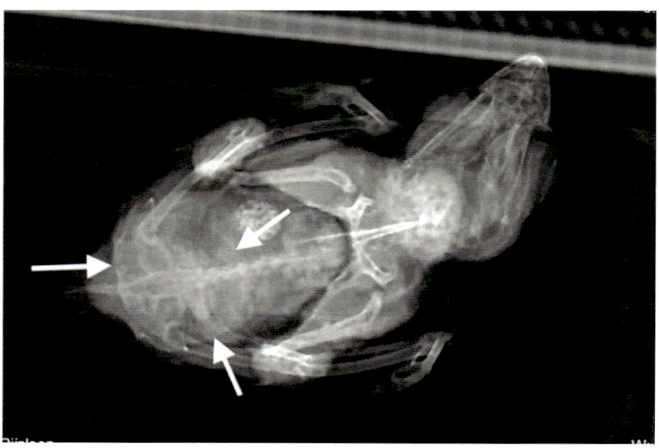

Fig. 28.4 X-ray of an awake, non-positioned bird with egg binding. A very thin egg shell is visible (arrows). Note the absence of hyperostosis in the long bones, indicating a disorder of calcium metabolism.

Note: The absence of a visible egg on X-ray does not rule out egg-related problems, as eggs are only clearly visible on X-rays after calcification of the shell. If egg-related problems are suspected, X-rays can be repeated 12–24 hours after calcium/vitamin D supplementation (see the next section), preferably administered IM. Ultrasound imaging can be useful to detect eggs without calcified shells.

THERAPY

Calcium

A calcium deficiency is a common cause of egg binding. Most birds with egg binding benefit from calcium supplementation. Therefore, start with calcium (boro) gluconate IM.

The further treatment depends on the situation:

- *Situation 1*: Egg binding with egg visible in the cloaca
- *Situation 2*: Egg binding with severe dyspnea or leg paralysis
- *Situation 3*: Other situations (most frequent)

Situation 1: Egg binding with egg visible in the cloaca (**Fig. 28.5**)

This condition mainly occurs in smaller species. In this condition, birds can be severely weakened due to the complete obstruction of the cloaca (Figure 28.5). Often dry mucous membranes are adhered to the eggshell.

Fig. 28.5 Egg binding with egg visible in the cloaca.

General stabilization

Birds with egg binding are stabilized using heat, nutrition, fluid therapy and oxygen (see General stabilization of sick birds, see p. 23). Assisted feeding through a crop tube is indicated in birds that have been anorexic for longer periods, but should be postponed until after removal of the egg in other situations. Regurgitation and aspiration of crop contents could otherwise be the result of handling and possibly sedation necessary for removal of the egg (see the next section).

Analgesia

This is a very painful situation and painkillers are indicated. Because dehydration is likely, opioids (e.g., butorphanol) are preferred to NSAIDs in the acute phase. After stabilization, NSAIDs (e.g., meloxicam) can be used instead of or in addition to the opioids.

Management

Mucous membranes sticking to eggs should be moistened immediately with physiological saline solution. A small amount of lubricant can be applied. Try to gently push the moistened membranes off the egg with a wet cotton tip. When all mucous membranes are loosened from the egg, an attempt can be made to gently massage the egg out of the cloaca. Be careful not to cause trauma to the soft tissues and not to press on the coelom/abdomen. When gently removing the egg is not possible this way, the egg can be imploded (see Appendix 7, p. 207). Removing the eggshell from the cloaca with forceps, needle holder or mosquito pliers is usually easy after imploding, but care must be taken not to cause damage to the mucous membranes while extracting sharp pieces of eggshell. When imploding is not possible, the egg shell can be broken into pieces with a mosquito forceps.

Situation 2: Egg binding with severe dyspnea or leg paralysis

General stabilization

Birds with egg binding and dyspnea are stabilized using heat, nutrition, fluid therapy and oxygen (see General stabilization of sick birds, see p. 23). Assisted feeding through a crop tube is indicated in birds that have been anorexic for longer periods, but should be postponed until after removal of the egg in other situations. Regurgitation and aspiration of crop contents could otherwise be the result of handling and possibly sedation necessary for removal of the egg (see the next section).

Analgesia

This is a very painful situation and painkillers are indicated. Because dehydration is likely and pressure on the ureters and kidneys by the egg is possible, opioids (e.g., butorphanol) are preferred to NSAIDs in the acute phase. After stabilization, NSAIDs (e.g., meloxicam) can be used instead of or in addition to the opioids.

Management

In this situation, there is a high risk of complications, such as death or permanent nerve damage. The egg can, in most cases, be imploded (see Appendix 7, p. 207) to create extra space in the coelomic cavity and to reduce pressure on the nerves. Imploding eggs in anesthetized birds with severe clinical symptoms is not without risk. Successful imploding leads to a rapid relief in most cases.

Note: Chronic egg binding, where eggs have been in the shell gland for longer periods can lead to thickened egg shells. Imploding eggs with extremely thick shells can be impossible and efforts to do so can result in serious complications.

After imploding, anorectic birds should be force fed (see Nutritional support, p. 33). A broad-spectrum antibiotic (e.g., amoxicillin/clavulanic acid) is indicated because of possible lesions in the reproductive tract due to the imploding and the risk of infection.

After imploding eggs and stabilizing, eggshells can be laid spontaneously within 1 or 2 days. If that does not happen or if the patient's general condition deteriorates, removal of the eggshell pieces through the cloaca or surgically by means of coeliotomy is indicated. Birds can be referred to an avian specialist for this procedure.

Situation 3: Other situations

General stabilization

In cases of egg binding without severe dyspnea, paralysis of legs and/or visibility of an egg in the cloaca, patients should first be stabilized using fluid therapy, heat, oxygen and nutrition (see General stabilization of sick birds, p. 23).

Analgesia

Egg binding is a painful situation and painkillers are indicated. In birds without dehydration, NSAIDs (e.g., meloxicam) can be used. In dehydrated birds, opioids (e.g., butorphanol) are preferred to NSAIDs in the acute phase. After stabilization, NSAIDs (e.g., meloxicam) can be used instead of or in addition to the opioids.

Stimulation of oviposition

Prostaglandin E2 gel (dinoprostone) administered into the cloaca is used to induce relaxation of the sphincter and stimulate contractions. In healthy, stabilized patients with sufficient calcium, application of prostaglandin E2 gel often results in rapid egg laying. Prostaglandin E2 gel is contraindicated when eggs are adhered to mucous membranes, ectopic eggs (eggs free in the coelom), misshapen eggs, eggs with an irregular shell surface (common in eggs with an extremely thick shell, see **Figs. 28.6** and **28.7**) or in case of obstruction.

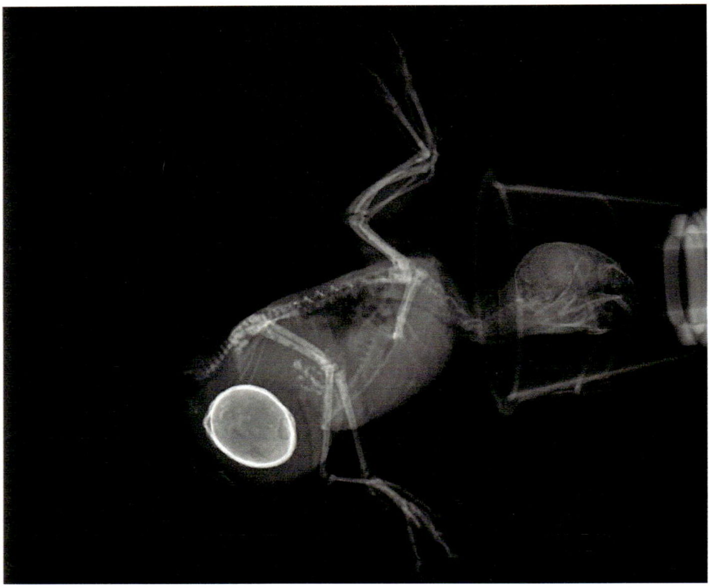

Fig. 28.6 X-ray of a lovebird showing an egg with an irregular surface and abnormal thickness of the shell. Imploding and prostaglandin E2 gel should not be tried because of the risk of complications and very unlikely positive effect.

Fig. 28.7 Egg with an irregular surface (as visible on the X-ray in Fig. 28.6), removed by coeliotomy.

Minimize stress

Stress slows down the laying process. Birds with egg binding are placed in a warm and quiet area.

If this initial approach does not result in laying, the calcium and prostaglandin E2 treatment can be repeated after six hours. When general stabilization, calcium injections, prostaglandin E2 gel and minimizing stress do not result in successful oviposition, invasive techniques for egg removal are mandatory.

Invasive techniques

Caudally located, palpable eggs can in most cases first be imploded (see Appendix 7, p. 207). Although spontaneous laying of eggshells after imploding can happen within 1 or 2 days, immediate cloacal removal of the eggshell parts is preferable in stable birds to minimize the risk of complications. When imploding and/or cloacal removal is impossible, eggs or eggshells should be removed by means of surgery (coeliotomy).

Note: These are delicate operations and require experience. Birds can be referred to an avian specialist for these invasive techniques.

29 *Dyspnea*

Symptoms of dyspnea can include rapid breathing (tachypnea), labored breathing, tail bobbing, cyanosis, open beak breathing and abnormal breathing sounds (stridor/stertor).

In birds, shortness of breath can be caused by disorders of the respiratory system (such as infection, hypersensitivity reaction, neoplasia or blockage of the trachea), cardiovascular system, or space-occupying lesions (such as free fluid in the coelom/abdomen, **Fig. 29.1**, eggs, neoplasia or enlarged organs) compressing the air sacs.

Fig. 29.1 Ascites can cause severe dyspnea. Fluids can be removed from the coelom by means of paracentesis for temporary stabilization of the patient and for diagnostic purposes.

Handling of dyspneic birds should be minimized and done with as little pressure on the body as possible to prevent restricting respiratory movements.

DOI: 10.1201/9781003308270-32

ADVICE FOR FIRST AID AT HOME

Owners should try to keep the bird calm by placing it in a quiet area with dimmed lights.

EMERGENCY CARE BY A VETERINARY PROFESSIONAL

Note: Dyspnea can be a sign of the zoonotic disease psittacosis (see Appendix 11, p. 225).

In case of dyspnea, physical examination is first aimed at differentiating between primary disorders of the respiratory system or disorders outside the respiratory system causing respiratory compromise, in most cases by compression of the caudal air sacs. X-rays (see Appendix 3, p. 179), ultrasound or computed tomography (CT) scans are very useful.

GENERAL STABILIZATION

OXYGEN

Dyspneic birds should first be stabilized in an oxygen cage. Air humidity in the oxygen cage should be sufficient (40–60%) when birds stay in for longer periods, as excessively dry air can lead to further respiratory problems.

NUTRITION AND FLUID THERAPY

In many cases, severely dyspneic birds are dehydrated and malnourished. Fluid therapy (see p. 27) is indicated, as is administering food through a feeding tube (see Nutritional support, p. 33), unless birds start eating immediately by themselves after being placed in the oxygen cage. Severely weakened birds should be fed after a short period of stabilization in the oxygen cage.

Diagnosing the cause of dyspnea can be quite challenging. Birds stabilized during the emergency consultation can be referred to an avian specialist for further diagnostics and treatment.

PARACENTESIS

In case of severe ascites, fluids can be removed from the coelom by means of paracentesis for temporary stabilization of the patient and for diagnostic purposes.

A number of specific cases are discussed here:

- Respiratory infections
- Pulmonary hypersensitivity reactions
- Tracheal obstruction
- Inhalation intoxication

RESPIRATORY INFECTIONS

Respiratory infections can be caused by bacteria, fungi (for example aspergillosis, **Fig. 29.2**), viruses and parasites. The clinical signs depend on the causative agent and the part of the respiratory system involved in the infection. Signs of upper respiratory tract infections can include abnormal breathing sounds, nasal discharge, conjunctivitis, swelling of the face, sneezing, coughing, shaking of the head and scratching or rubbing the head. Lower respiratory tract infections can affect the trachea, syrinx, bronchi, lungs and/or air sacs. Signs of infections of the trachea and bronchi include coughing and (sudden) dyspnea with a loud stridor. Signs of lung infection include dyspnea without a loud stridor, exercise intolerance, and general signs of illness. Infections of the air sacs don't necessarily cause severe dyspnea in the early stages of disease. Instead, weight loss and symptoms of general illness are more common. Naturally, infections are not necessarily limited to one part of the respiratory tract, and several areas may be affected. Thus, a stridor does not rule out pneumonia, and sudden dyspnea does not rule out air sac involvement.

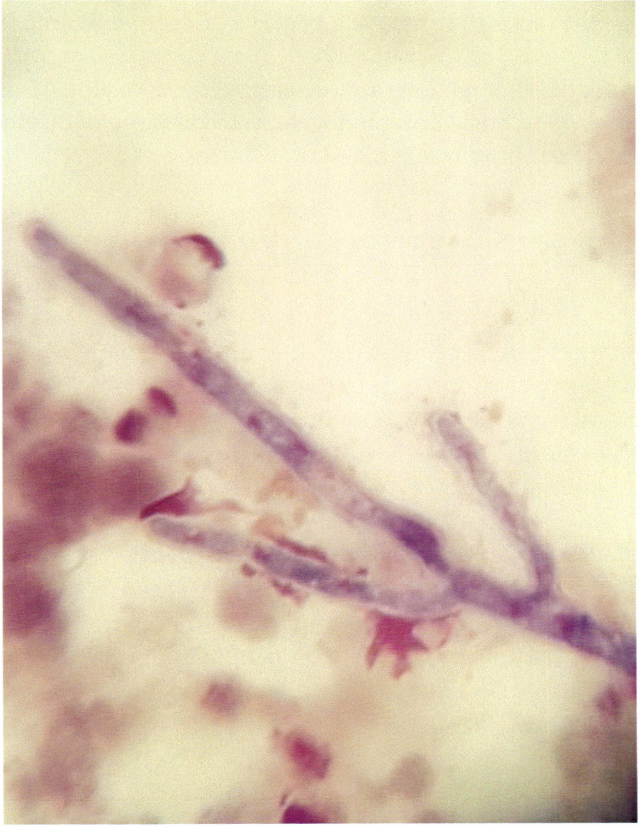

Fig. 29.2 Microscopic view of aspergillus. Aspergillosis is a fungal infection that can affect the entire respiratory system.

EMERGENCY CARE BY A VETERINARY PROFESSIONAL

In case of nasal or eye discharge, smears of exudate are made for microscopic examination and samples are taken for bacteriological culturing and susceptibility testing. Samples of exudate, choana and cloaca are taken with dry swabs for polymerase chain reaction (PCR) testing for *Chlamydia psittaci*.

Bacterial infections are treated with antibiotics (e.g., doxycycline or amoxicillin/clavulanic acid). In cases of bacteriological culturing and susceptibility testing, treatment with antibiotics is started pending culture results.

Fungal infections are treated with antifungal agents (e.g., itraconazole, voriconazole, terbinafine, clotrimazole or amphotericin B).

Note: Ideally, a diagnosis will be made before starting therapy. Unfortunately, this is not always possible during emergency consultations in general veterinary clinics. Although dyspnea can be caused by many other causes than bacterial infections, starting antibiotic treatment can be considered in cases of dyspneic birds when no other cause can be identified during the emergency consultation.

NSAIDs (e.g., meloxicam) can have a positive effect due to anti-inflammatory effects.

Many respiratory tract infections are secondary to underlying disorders, such as inappropriate diets or husbandry. If not corrected, relapse of dyspnea or other physical or mental health problems are likely to occur.

PULMONARY HYPERSENSITIVITY REACTIONS

Pulmonary hypersensitivity reactions (allergic asthma attacks) can cause sudden and severe dyspnea. Hypersensitivity reactions are particularly common in macaws living indoors in dusty environments or together with other psittacines (particularly cockatoos). Clinical signs include labored breathing, breathing with an open beak and abnormal breathing sounds. Usually, no further abnormalities are found in the general physical examination.

EMERGENCY CARE BY A VETERINARY PROFESSIONAL

Birds with pulmonary hypersensitivity reactions are placed in an oxygen cage. A bronchodilator (e.g., salbutamol) is administered to immediately relieve signs of the hypersensitivity reaction by means of bronchodilation. NSAIDs (e.g., meloxicam) can also have a positive effect due to anti-inflammatory effects.

TRACHEAL OBSTRUCTION

Obstruction of the trachea can cause life-threatening dyspnea with, in most cases, labored open-beak breathing with a loud stridor. Tracheal obstructions are usually caused by fungal infections or aspiration of seeds/seed hulls, but can also be caused by, for example, trauma, stricture formation (for example as a postoperative complication after intubation), neoplasia and aspiration of medication, formula or crop contents.

EMERGENCY CARE BY VETERINARY PROFESSIONAL

If placement of birds with signs of a tracheal obstruction in an oxygen cage does not rapidly lead to significant reduction of symptoms of dyspnea, placement of an air sac tube (see Appendix 6, p. 201) is indicated for stabilization. Birds can breathe effectively through air sac tubes. **Fig. 29.3**.

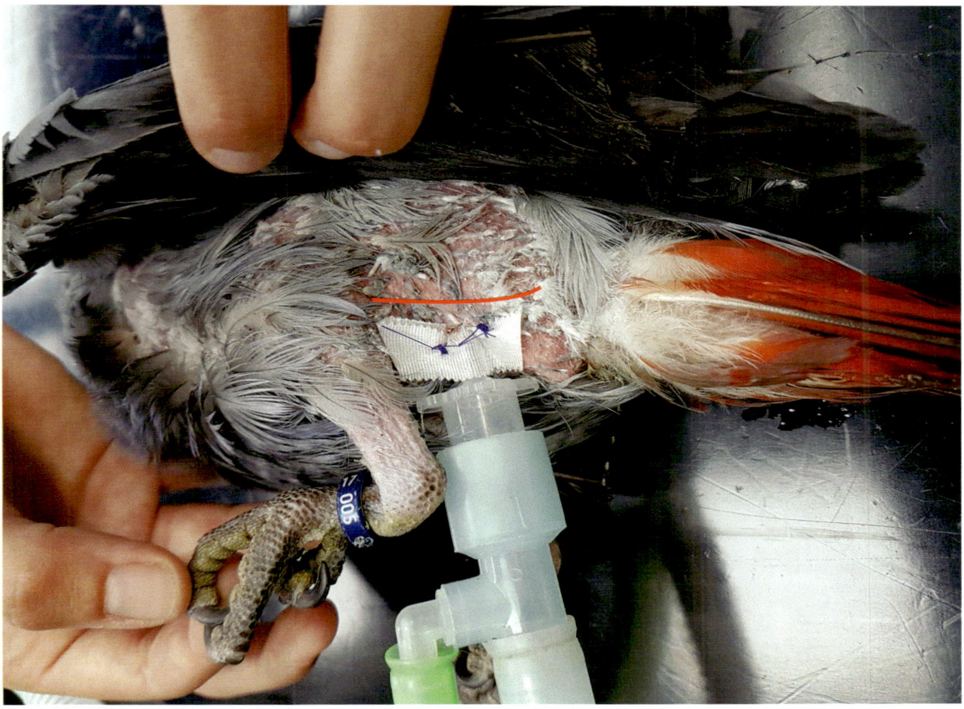

Fig. 29.3 Anesthetized African grey parrot breathing through an air sac tube.

Intubation of air sacs provides instant relief in birds with trachea obstruction, unless there is further pathology of the lower airways (lung or air sac disease). Because air sac intubation does not solve the cause of dyspnea and can only be left in place for a few days, birds should be referred to an avian specialist for further diagnostics (for example endoscopy of the trachea, X-rays and/or CT scan) and treatment after stabilization.

INHALATION INTOXICATION

Inhalation of toxic fumes, gases or smoke can lead to dyspnea by intoxication through the respiratory tract (see Inhalation intoxication, p. 71).

30 *Falling, abnormal stances and abnormal movements*

FALLING, abnormal stances and abnormal movements are not signs of specific diseases, but can be caused by numerous things, including orthopedic abnormalities like arthritis (**Fig. 30.1**), fractures or luxations (see Abnormal position of limbs: Fractures and luxations, p. 131), cardiovascular diseases, neurological problems (e.g., trauma, infections of the central nervous system, neoplasia, stroke or intoxications [see p. 71]), egg binding (see p. 105), hypocalcemia (see Appendix 12, p. 227), gout (**Fig. 30.2**) or pododermatitis/bumblefoot (**Fig. 30.3**).

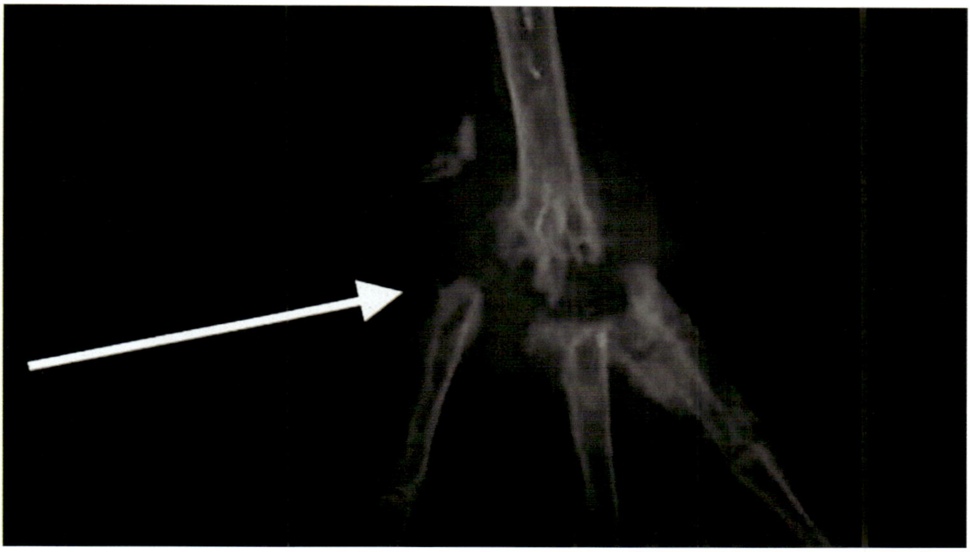

Fig. 30.1 Septic arthritis with osteolysis of the metatarsophalangeal joint (arrow) in a mallard.

DOI: 10.1201/9781003308270-33

Fig. 30.2 Gout as a result of kidney failure. Note the white periarticular swellings.

Fig. 30.3 Pododermatitis.

ADVICE FOR FIRST AID AT HOME

Trauma should be prevented by removing high perches. One perch should be left just above the bottom of the cage. The bottom should be covered with a soft substrate or towel. Birds that keep climbing high on the bars of their cages should be placed in a small cage or transport box.

EMERGENCY CARE BY A VETERINARY PROFESSIONAL

Note: Neurological problems can be a sign of the zoonotic disease psittacosis (see Appendix 11, p. 225).

Anamnesis, physical examination and further diagnostics are aimed at diagnosing or ruling out intoxications (see p. 71), egg binding (see p. 105), hypocalcemia (see Appendix 12, p. 227) and orthopedic conditions like arthritis, fractures and dislocations (see Abnormal position of limbs, p. 131), gout, neoplasia and pododermatitis.

GENERAL STABILIZATION

Sick birds are stabilized using heat, nutrition, fluid therapy and oxygen (see General stabilization of sick birds, see p. 23). Oxygen is indicated in case of dyspnea and when cardiovascular disease cannot be ruled out.

Further treatment depends on the cause of the problem (see specific chapters and appendix mentioned previously).

31 *Paralysis*

Paralysis is the loss of muscle function in a part of the body. Paralysis is caused by problems of the nervous system, for example trauma of the spinal cord or peripheral nerves, stroke, intoxication, neoplasms of the nervous system or pressure on nerves by eggs or neoplasms/tumors in the coelom/abdomen.

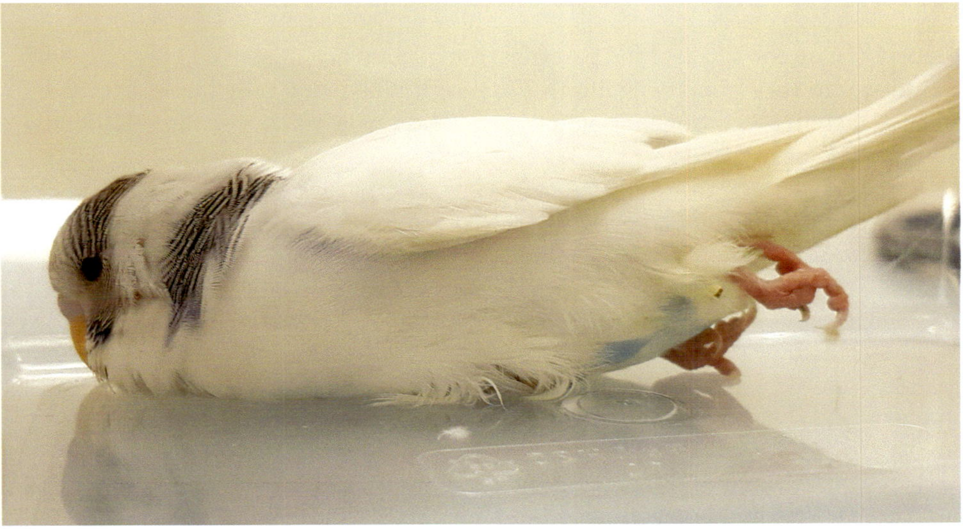

Fig. 31.1 Budgerigar with bilateral paralysis of the legs.

A common cause of bilateral paralysis (**Fig. 31.1**) in non-geriatric psittacines in captivity is lead poisoning. In free-ranging birds, trauma to the back and spinal cord is a notorious cause of bilateral paralysis. Botulism frequently leads to paralysis in waterfowl. Marek's disease frequently leads to paralysis in young chickens.

EMERGENCY CARE BY A VETERINARY PROFESSIONAL

Medical history taking, physical examination and further diagnostics are initially aimed at diagnosing or ruling out lead poisoning (see p. 73), egg binding (see p. 105) and trauma (see Abnormal position of limbs: Fractures and luxations, p. 131).

Note: Most birds don't react clearly to painful stimuli during physical examination. Loss of leg function caused by fractures or luxations can be mistaken for paralysis.

DOI: 10.1201/9781003308270-34

RADIOLOGY

X-ray examination (see Appendix 3, p. 179) or computed tomography (CT-scan) is indicated in cases of paralysis to check for fractures and luxations (see Abnormal position of limbs: Fractures and luxations, p. 131), metal particles in the gastrointestinal tract, calcified eggs or space-occupying lesions (e.g., tumor) in the coelom.

GENERAL STABILIZATION

Depending on the cause of the paralysis, the general health status of patients with paralysis can vary from being life-threateningly ill to being fully stable. Provide general stabilization by means of nutrition, fluids, heat and oxygen (see General stabilization of sick birds, p. 23), if necessary.

Note: Because ruling out lead poisoning is often not possible during emergency consultations, starting treatment for lead intoxication in non-geriatric psittacines in captivity with bilateral leg paralysis could be considered in case no other cause can be determined.

ANALGESIA

If there is evidence of trauma, NSAIDs (for example, meloxicam) are indicated for analgesia and to reduce swelling around the nerves. Additional analgesics like opioids (e.g., butorphanol, tramadol) and/or gabapentin are indicated in cases of severe (neurological) pain.

32 *Abnormal eye or closed eyelids (inability or unwillingness to open the eye)*

EYELIDS can be closed due to pain (**Fig. 32.1**), swelling or exudate. Examples of disorders causing the eyes to be closed are head trauma, injuries or infections of the cornea or eyelids, foreign objects behind the eyelids, intra-ocular problems (for example uveitis, an inflammatory reaction of the inner parts of the eye) or tumors. Disorders of the cornea and intra-ocular diseases can also change the appearance of the eye.

Fig. 32.1 Blepharospasm in a pyrrhura.

ADVICE FOR FIRST AID AT HOME

If the eyelids are stuck together by dried exudate, owners can try to soften and remove the crust by repeatedly applying drops of saline solution and gently wiping with a cotton swab. Owners should not touch the eye itself, only the outside parts of the eyelids. Touching the cornea might lead to serious complications.

EMERGENCY CARE BY A VETERINARY PROFESSIONAL

Note: Conjunctivitis can be a sign of the zoonotic disease psittacosis (see Appendix 11, p. 225).

Closed eyes must be opened for examination of the eye and for prevention of further damage to the cornea in case of infections or foreign bodies. Anesthesia is frequently necessary or useful in cases of painful disorders of the eyes.

DOI: 10.1201/9781003308270-35

In case the eyelids are stuck, drops of saline solution can be applied to soften dried exudate. Next, the eyelids can be gently opened using wet cotton tip applicators without putting any pressure on the cornea.

The eyes are examined just like in other companion animals. Fluorescein staining can be used to reveal corneal damage.

In case of inflammation, samples are taken for bacteriological culturing and susceptibility testing. Samples of exudate, choana and cloaca can be taken with dry swabs for polymerase chain reaction (PCR) testing for *Chlamydia psittaci*.

In case of severe sinusitis with swelling and fluctuation of the infraorbital sinus (**Fig. 32.2**), the sinus can be punctured and fluid contents can be aspirated for instant relief, microscopic examination and bacteriological/PCR testing. In cases of solid contents, the infraorbital sinus can be opened under anesthesia with a small incision for removal of caseous debris/exudate or for biopsy in cases of tissue masses.

Fig. 32.2 Eye closed caused by swelling of the infraorbital sinus.

Foreign objects and exudate should be removed with care, preferably by flushing with saline solution.

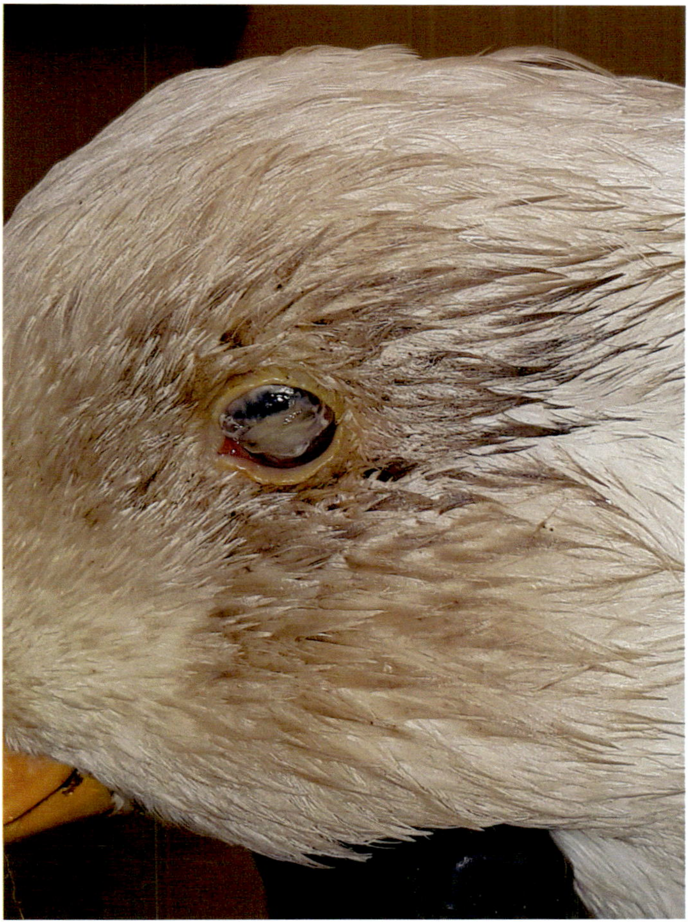

Fig. 32.3 Keratitis caused by a mycotic (Aspergillus) and bacterial infection in a Pekin duck.

Note: In case of severe keratitis, the thickened and damaged cornea might resemble stuck exudate ***(Fig. 32.3)****. Attempts to remove this tissue may result in corneal perforation and should therefore not be made initially.*

ANTIBIOTICS

In cases of bacterial conjunctivitis and keratitis, topically and/or systemically administered antibiotics are indicated. In cases of sinusitis or uveitis, systemically administered antibiotics are indicated.

ANTI-INFLAMMATORY ANALGESICS

Because of systemic uptake and serious side effects, corticosteroids are contraindicated. NSAIDs (e.g., meloxicam) are indicated for analgesia and anti-inflammatory effects in case of inflammation or other painful disorders.

Referral to an avian specialist/ophthalmologist is indicated in case of intra-ocular abnormalities.

33 *Abnormal position of limbs: Fractures and luxations*

ABNORMAL positions of legs and wings (**Fig. 33.1**) are, in most cases, due to pain, fractures, joint dislocations/luxations or paralysis (see Paralysis, p. 125). For anatomy of the avian skeleton, see Anatomy, p. 237.

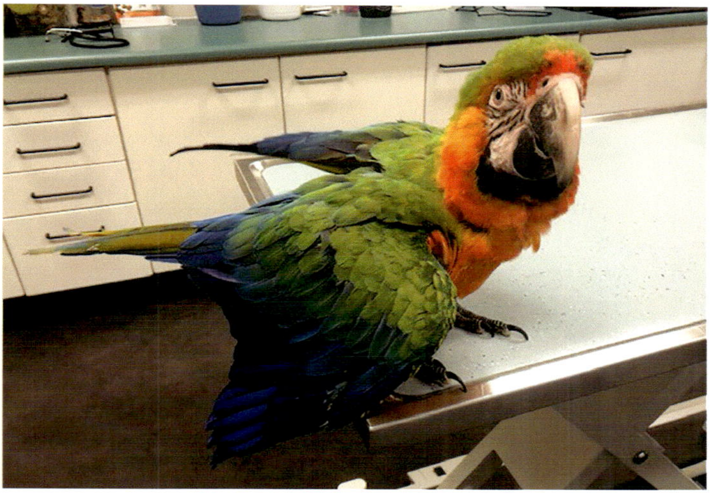

Fig. 33.1 Macaw with drooped wing indicating severe pain and/or orthopedic problems of the wing.

Fractures are common in birds. In free-ranging birds, fractures usually occur through collisions/trauma with major impact on healthy strong bones. In domestic birds, fractures regularly occur through trauma with relatively low impact on bones weakened by osteoporosis. The latter is usually due to a lack of calcium, vitamin D3 and UV-B light. Fractures are very painful (even though birds do not always show this) and can lead to significant blood loss through external or internal bleeding and infections if the skin is also damaged (open fracture).

Joint dislocation or luxation also occurs as a result of trauma. Knee and hock luxations often result from traction and torsion instead of high-velocity impact. Compared to fractures, joint dislocations are less often associated with other internal injuries and blood loss/big hematomas. Joint dislocations are painful and, in addition to damage to the joint itself, can lead to problems with blood circulation, skin and soft tissues.

DOI: 10.1201/9781003308270-36

ADVICE FOR FIRST AID AT HOME

Whether owners could and should do anything in case of abnormal position of limbs depends on what body part is affected and—especially in case of fractures—whether there is perforation of the skin at the site of the damaged bone.

As long as the skin at the fracture site is not damaged (closed fracture), there is no risk of infection. It is important to prevent further damage to the injured body part. Sharp edges from fractures can cause damage to the surrounding soft tissues and can perforate the skin from the inside. The less the damaged body part is moved, the better it is. Immobilizing and supportive bandages can help to prevent pain and complications. However, incorrectly applied bandages can cause major complications.

In closed fractures and joint dislocations/luxations of the legs, bandages should not be applied by inexperienced owners. To make sure that loading and moving of the injured leg is minimized, the injured bird should be placed in a small transport cage or, in cases of calm and tame birds, be gently held.

In closed fractures of wings (especially of the upper wing, resulting in a strong wing droop), the risk of skin perforation by sharp bone parts is significant. Immobilizing the injured wings reduces this risk. The easiest way for owners to do this is by applying a body bandage or by taping the wing tips together (see Appendix 8, p. 209).

If the skin is already damaged at the site of the fracture or luxation (open fracture or luxation) of the leg, protection against contamination and devitalization of the bone is crucial. Bacterial infection interferes with bone healing and carries a worse prognosis. Owners should put on sterile gloves to prevent contaminating the wound with bacteria from the hands and cover the wound with sterile gauze, preferably moistened with sterile physiological saline solution. A loose layer of elastic bandage can be applied to keep the gauze in place. In severe cases where bone fragments protrude from the wounds, keeping the gauze moist is beneficial for preventing drying and devitalization of the bone. To make sure that loading and moving of the injured leg is minimized, the injured bird should be placed in a small transport cage or, in case of calm and tame birds, be gently held.

In open fractures of wings without bone fragment protruding from the wounds, the risk of devitalization of the bone is small. First aid is as described previously for closed fractures, but with sterile gloves on to prevent contaminating the wound with bacteria from the hands. In open fractures of wings with bone fragments perforating the skin, protection against contamination and devitalization of the bone is crucial. Owners should put on sterile gloves and cover the wound with sterile gauze, moistened with sterile physiological saline solution. A body bandage should be applied to hold the gauze in place and to prevent the wing from being moved. Calm and tame birds can be gently held by the owner.

EMERGENCY CARE BY A VETERINARY PROFESSIONAL

Most fractures and dislocations can be determined by careful palpation and inspection (**Fig. 33.2**). However, soft tissue swelling, hematomas and coverage with muscle mass can make this challenging in some cases, especially in hip luxation or fractures of the proximal femur. Differentiating between luxations and fractures close to joints is not always possible by palpation.

Note: Most birds don't react clearly to painful stimuli during physical examination. Loss of leg function caused by fractures or luxations can be mistaken for paralysis.

Fig. 33.2 Elevation of the tip of the wing (in this case, of a European goldfinch) can be the result of a fracture of the coracoid.

RADIOLOGY

X-rays (see Appendix 3, p. 179) or computed tomography (CT scan) are needed for definitive diagnosis. In case of fractures, the optimal treatment method depends on the exact location and fracture classification.

Note: It is not always necessary to take X-rays or CT scans directly during the emergency consultation. In particular, when referral to an avian veterinarian experienced in orthopedics is possible after first aid treatment. X-rays or CT scans can be made by the specialized veterinarian instead of during the emergency consultation. This way, compromised birds are not exposed to unnecessary risks and optimal positioning for the X-rays is more likely.

Stabilization of the patient, adequate analgesia and prevention of infection have priority over treatment of the fracture or luxation.

STABILIZATION

Fluid therapy (see p. 27) is indicated in case of external or internal blood loss (including large hematomas). Injured wildlife can have been anorexic for prolonged periods and be severely debilitated (see Quick guide for stabilizing birds in case of severe dyspnea, debilitation and shock, p. 23).

ANALGESIA

Broken bones and joint dislocations are very painful. Analgesics are indicated (e.g., meloxicam and butorphanol or tramadol). In cases of significant blood loss (>1% of body weight), possibly leading to hypovolemia and decreased kidney function, NSAIDs should be used with care until circulation has been sufficiently restored.

FRACTURES

In open fractures, systemic antibiotic therapy (e.g., amoxicillin/clavulanic acid) must be started immediately. Foreign material (feathers, bedding, plant parts, etc.) present should be removed from the wound. This is done with sterile gloves and sterile instruments, even if the wound is already infected. Samples for bacteriological examination are taken if the wound is not fresh. The wound is then rinsed generously with physiological saline solution at body temperature. Take care not to flush fluids into the medullary cavity of pneumatized bones like the humerus and femur, as this can lead to drowning or spreading of infections into the respiratory system. The wound is disinfected once with diluted povidone iodine

(1% solution). Bone parts protruding through the skin must be brought back under the skin, as all bone must be covered by soft tissues to prevent devitalization of the bone tissue. Devitalized wound edges are debrided. The skin is closed with monofilament sutures.

STABILIZATION OF FRACTURE PARTS

Immobilizing and supportive bandages or splints can help to reduce the risk of complications and unnecessary pain. See Appendix 8, p. 209.

Note: In certain cases, correctly applied (splint) bandages may be the best definitive treatment option. Examples are fractures of bones with a medullary cavity of less than 0.6 mm (so all bones in very small species or small bones in small species), fractures of the shoulder girdle, single fractures of the carpometacarpus, fractures of the tarsometatarsus in most species and single fractures of the radius or ulna (not radius and ulna) without dislocation. The alignment of bone fragments after applying the bandage should be checked if this is meant as the permanent treatment instead of temporary immobilization.

Most fractures are best permanently treated with surgery instead of immobilization with bandages or splints. Surgery offers the best changes of a functional and pain-free recovery. For this, birds can be referred to an avian veterinarian skilled in orthopedics.

Femur

Due to the anatomy of birds, it is usually difficult to stabilize a femoral fracture with a bandage. Fortunately, the bone in the thigh is usually sufficiently protected by the soft tissues to prevent skin damage from within. In the case of a femoral fracture, no support bandage is applied during the emergency consultation.

Lower leg

Fractures distal to the knee (tibiotarsus, **Fig. 33.3**; tarsometatarsus, **Fig. 33.4**, phalanx bones) can be more or less stabilized by applying bandages or splits.

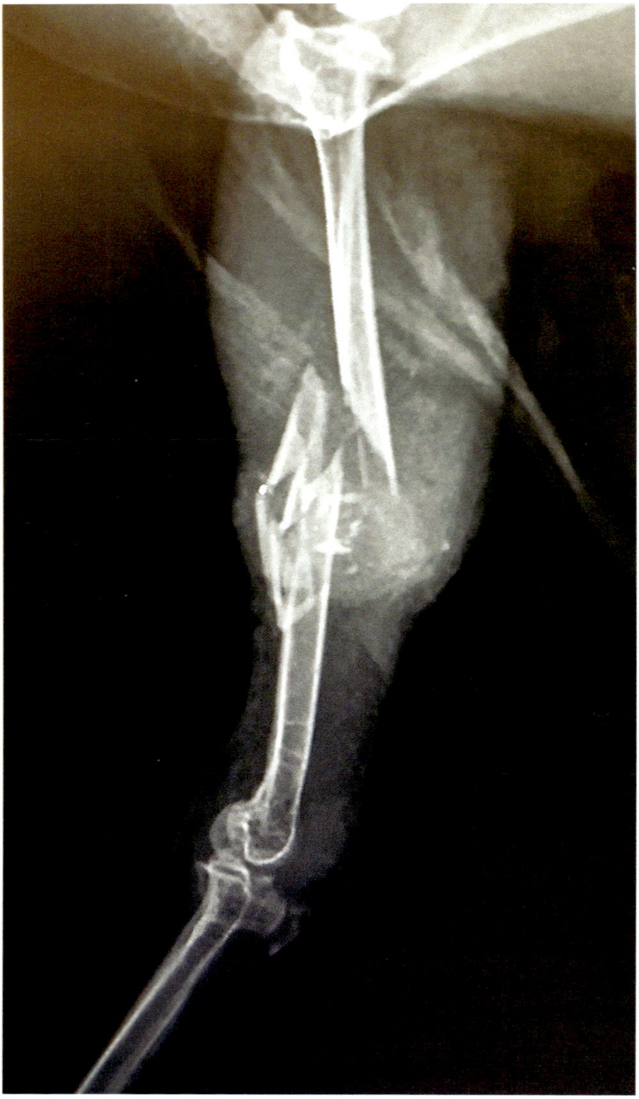

Fig. 33.3 Comminuted fracture of the tibiotarsus caused by a gun shot.

For larger birds, Robert Jones bandages can be applied, consisting of a thin layer of padding covered with an elastic self-adhesive bandage or tape. A hard splint made from wood, plastic, K-wire, synthetic resin, etc., can be incorporated. Immobilizing the joints proximal and distal to the fracture gives the best results, but in case of tibiotarsal fractures, this is not always possible, as the knee and medial thigh lie against the body wall. Bandages should not be made too heavy or bulky, as this can lead to complications. Joints should be immobilized in normally bent resting position, so that the leg with the splint is no longer than the other leg when the bird is standing; this makes a huge difference in discomfort for the patient and reduces the risk of complications due to an abnormal load on the leg.

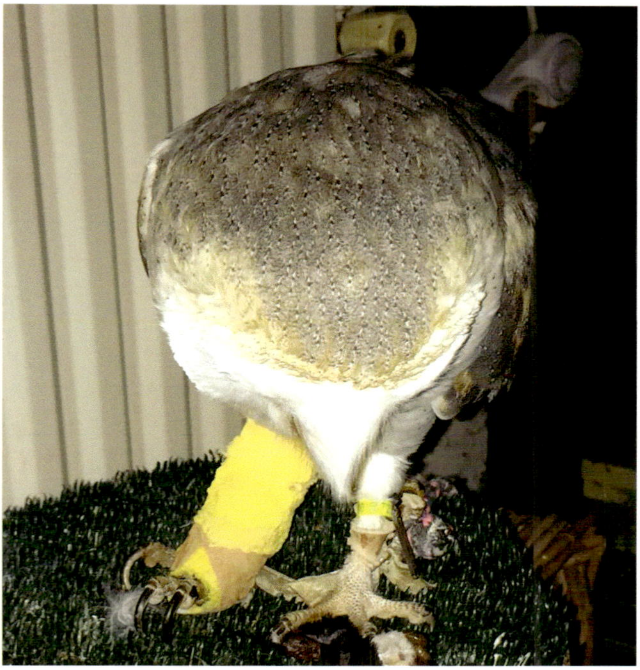

Fig. 33.4 Barn owl with fracture of the tarsometatarsus stabilized by a Robert Jones bandage, including the tarsometatarso-phalangeal joints and the hock joint.

In small birds, the bones of the lower leg can be stabilized with a tape splint.

Wing

Fractures distal to the elbow (radius, ulna or carpometacarpal bones, **Fig. 33.5**), can be stabilized by a figure-of-eight bandage.

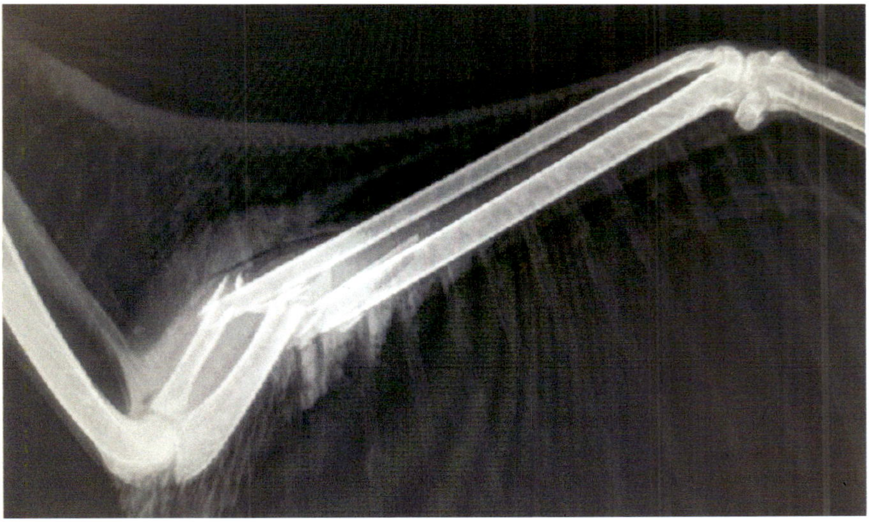

Fig. 33.5 Fractured radius and ulna in a mute swan.

The humerus (**Fig. 33.6**) cannot be stabilized by a figure-of-eight bandage. In case of a fractures of the humerus, a body wrap can be applied or the wing tips can be taped together for stabilization. Taping the wing tips together is less effective than a body wrap, but can be useful in emergency situations.

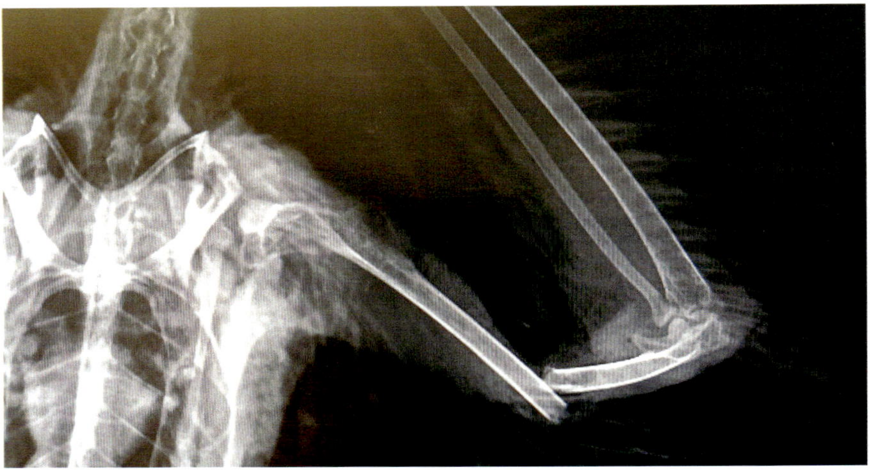

Fig. 33.6 Fractured humerus in a seagull.

JOINT DISLOCATION/LUXATION

EMERGENCY CARE BY VETERINARY PROFESSIONAL

Stabilization

Fluid therapy (see p. 27) is indicated in case of external or internal blood loss (including large hematomas). Injured wildlife can have been anorexic for prolonged periods and be severely debilitated (see Quick guide for stabilizing birds in case of severe dyspnea, debilitation and shock, p. 23).

Analgesia

Joint dislocations are very painful and analgesics (e.g., meloxicam and butorphanol or tramadol) are indicated.

Reduction

It is important to reduce dislocations for minimizing secondary problems as soon as possible after general stabilization of the patient.

Note: Closed reduction of acute dislocations can be challenging or even impossible. Care should be taken not to cause extra damage like fractures. If referral to an avian specialist is possible, this is preferred over attempts to reduce challenging luxations by veterinarians without experience.

Reduction is painful and difficult with tensed muscles and should therefore be done under general anesthesia (e.g., isoflurane/sevoflurane) with adequate analgesia (e.g., meloxicam and butorphanol).

Immobilizing and supportive bandages or splints can help to reduce the risk reluxation, see Appendix 8, p. 209.

Shoulder luxation

Luxations of the shoulder are very unstable and can involve fractures of the humerus or shoulder girdle. Reduction is relatively easy, but reluxation is common. The shoulder joint can be stabilized by applying a figure-of-eight bandage in combination with a body wrap. Specialist referral of bigger species is indicated, as open reduction and surgical stabilization is sometimes possible.

Elbow luxation

The elbow is flexed and radius and ulna are endorotated. Pressure is applied on the dorsal side of the proximal radius close to the joint to push the head of the radius back into place. Then, the elbow is extended again, hopefully resulting in normal alignment of the ulna and the rest of the joint. The elbow joint is stabilized by applying a figure-of-eight bandage in combination with a body wrap.

Carpometacarpal, metacarpophalangeal and interphalangeal joint (of the wing) luxation

After closed reduction, joints can be stabilized by applying a figure-of-eight bandage.

Coxofemoral joint/hip luxation

Luxation in craniodorsal directions is most common. The leg is exorotated and the proximal femur is pushed toward the acetabulum. Endorotating the leg while applying some pressure medially on the trochanter can result in reduction of the luxation. After successful reduction, a body wrap including the affected leg (with bent knee and hock) can be applied to prevent exorotation and reluxation. Specialist referral is indicated for reduction and surgical stabilization if reduction is not successful.

Femorotibial joint/knee luxation

Luxations of the knee can generally not be reduced and successfully stabilized without surgery, because of the instability of this joint after rupture of the ligaments. Specialist referral is indicated for open reduction and surgical stabilization.

Tibiotarso-tarsometatarsal joint/hock luxation

Closed reduction of acute luxation is usually not difficult. Reluxation is common, but applying immobilizing bandages (Robert Jones bandages in bigger species or tape splints for small species) can be very successful. Specialist referral is indicated for reduction and surgical stabilization if closed reduction and stabilization is not successful.

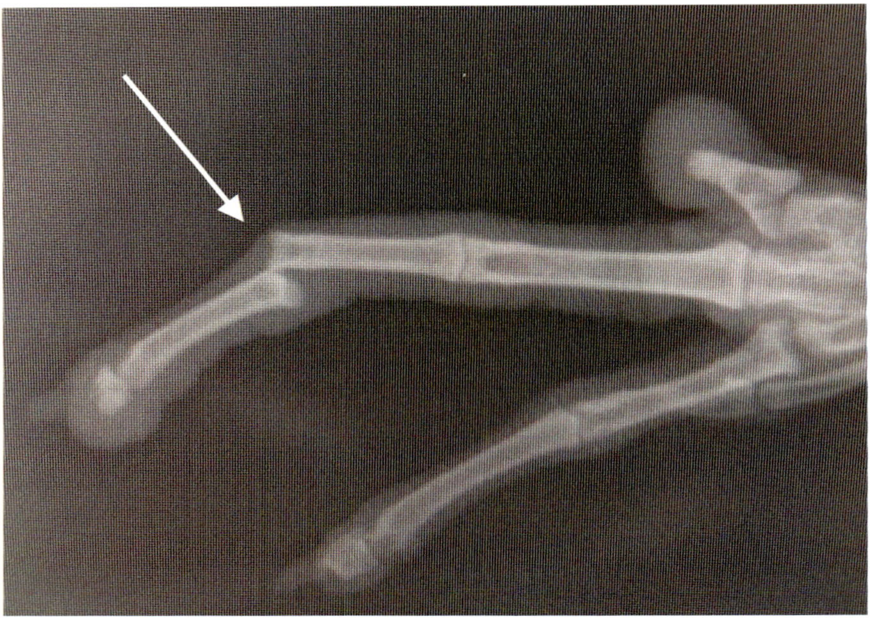

Fig. 33.7 Dislocation of interphalangeal joint (arrow) of a grey crowned crane.

Metatarsophalangeal and interphalangeal joint (of the leg) luxation

After closed reduction, joints can be stabilized by applying bandages. In case of luxation of metatarsophalangeal joints, bandages should include the whole foot (up to the proximal tarsometatarsus and splints can be included). In case of luxation of distal interphalangeal joints (**Fig. 33.7**), tape splints made of circular adhesive tape can be sufficient. In more proximal luxations, bandaging the whole foot might be necessary.

34 *Maxillary hyperextension/ palatine bone luxation*

THE ANATOMY of the avian skull and jaws is more complex than the mammalian skull. Psittacine birds possess a joint between the upper jaw and frontal bone, enabling movement of the upper beak in addition to the movement of the lower beak. Movement of the palatine bones coincides with movement of the upper jaw. Trauma or biting on big very hard objects can result in hyperextension of the upper beak and luxation of the palatine bones, which get stuck on the interorbital septum, thereby preventing closing of the beak (**Fig. 34.1**). Although spontaneous resolution of the hyperextension is possible, in many cases this does not happen. Birds with hyperextension of the maxilla cannot manipulate objects or food items, resulting in a decreased intake of nutrients.

Fig. 34.1 Blue-and-yellow macaw with maxillary hyperextension/palatine bone luxation.

EMERGENCY CARE BY A VETERINARY PROFESSIONAL

GENERAL STABILIZATION

Depending on the chronicity of the disorder, general stabilization with fluid therapy, nutritional support and heat can be indicated (see General stabilization of sick birds, p. 23).

DOI: 10.1201/9781003308270-37

ANALGESIA

Analgesics (e.g., meloxicam and butorphanol) are indicated because of the painful disorder and treatment.

MANAGEMENT OF THE HYPEREXTENSION

For normal movement of the maxilla, the palatine bones should be freed from the septum. Under general anesthesia (for example with isoflurane/sevoflurane) with sufficient analgesia (e.g., meloxicam and butorphanol), a Kirschner wire is inserted through the infraorbital sinus. The introduction site is cranial to the eyes and just caudal to the commissure of the beak. A soft depression can be palpated, as the infraorbital sinus is not covered with bone in this spot. The Kirschner wire should at least be inserted to just over the midline, but can also be introduced transversely across both infraorbital sinuses all the way through the skin on the contralateral side. A correctly placed Kirschner wire is located just dorsal to the palatine bones (**Figs. 34.2** and **34.3**).

To free the palatine bones from the septum, the maxilla is hyperextended a bit more and then the Kirschner wire is pushed ventrally. Successful reduction immediately results in normal mobility of the upper jaw (**Fig. 34.4**).

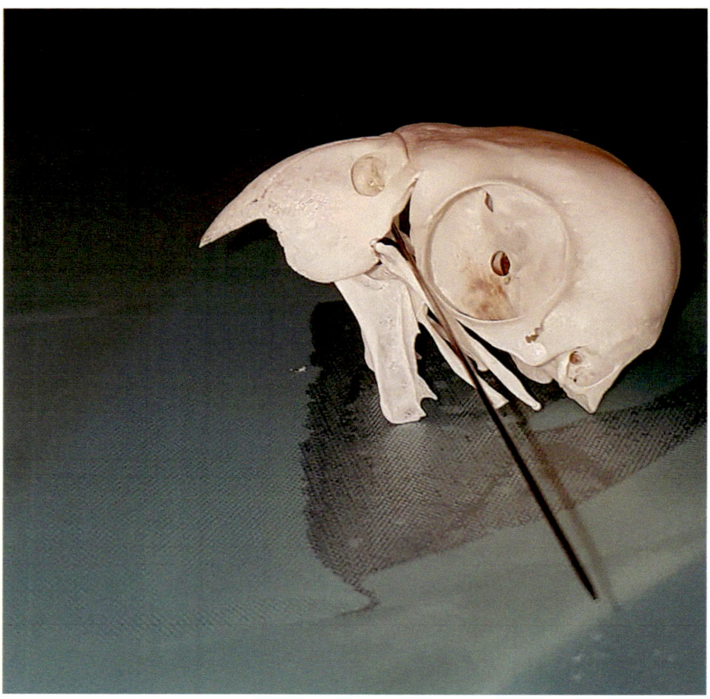

Fig. 34.2 Placement of Kirschner wire through the infraorbital sinuses dorsal to the palatine bones.

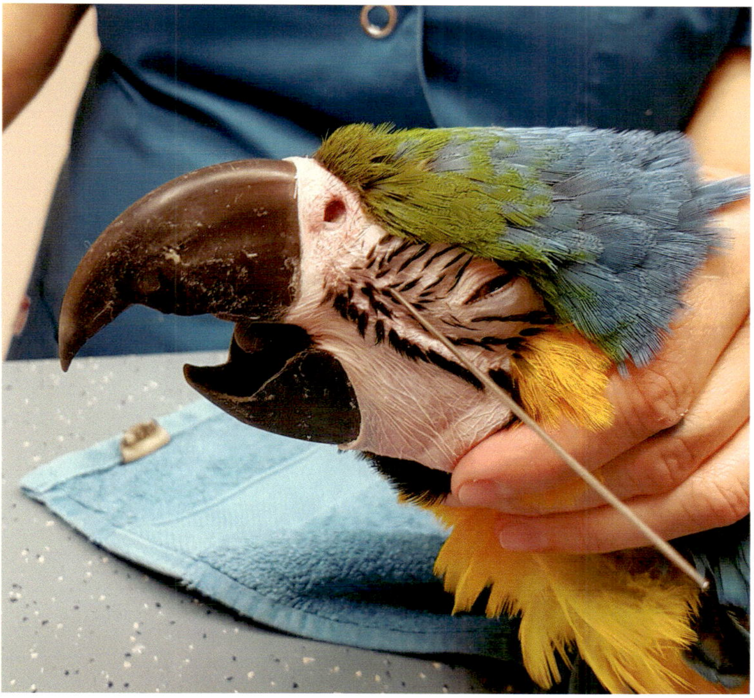

Fig. 34.3 Placement of Kirschner wire through the infraorbital sinuses of an anesthetized blue-and-yellow macaw with maxillary hyperextension/palatine bone luxation.

Fig. 34.4 Successful reduction immediately results in normal closure of the upper jaw.

AFTERCARE

To prevent recurrence, birds should not be allowed to climb or bite on hard (food) objects for 1–2 weeks. Small or soft food items should be offered and analgesics (e.g., meloxicam) should be continued for 1–2 weeks.

If recurrence repeatedly happens directly after the procedure, sutures can be placed around the ventral orbital ring (extreme care should be taken to prevent damaging the eyes) and the jugular bone. Keeping the beak closed with tape for a few hours can also prevent direct recurrence, but it, of course, also prevents the bird from eating and drinking and causes a fatal risk in case of vomiting/regurgitation.

35 *Abnormal droppings*

AVIAN droppings normally consist of feces, uric acid crystals (urates) and watery urine. The color and consistency of the feces can vary with the diet. The feces of most birds is quite firm and has a green or brown color. Those of birds of prey and owls often have a darker color. Urates from fresh droppings should always be white (urates in older droppings can be colored due to contact with feces).

This chapter discusses the following types of abnormal droppings:

1. Significantly decreased amount of feces.
2. Black stools.
3. Pink urates.
4. Yellow/green urates.
5. Fresh blood.
6. Diarrhea.

SIGNIFICANTLY DECREASED AMOUNT OF FECES

When the amount of feces is significantly decreased (**Fig. 35.1**), droppings will mainly consist of urates and watery urine. A small amount of dark green bile may be present. In most cases, a decrease in the amount of feces is caused by decreased food intake. Decreased intake of food can be the result of, for example, absence of food, decreased appetite as a result of illness or inability to process food items due to physical problems. Besides decreased food intake, a decreased amount of feces can also be caused by vomiting (see p. 89), crop stasis (see p. 95), constipation, egg binding (see p. 105), reduced intestinal motility or intestinal obstruction.

ADVICE FOR FIRST AID AT HOME

Owners should offer favorite food items and try to stimulate/encourage the bird to eat. Intake of energy is most important, and initially for a short term, the exact nutrient composition of the food is of minor importance as long as the food items being offered are not harmful. For example, psittacines can be offered sunflower seeds or spray millet. When tame birds don't eat at all but are well awake, liquid food may be administered into the beak with a spoon or syringe (from the side of the beak and not all the way back in the throat, as this can lead to aspiration and death). In general, 2 ml of liquid diet per 100 grams of body weight is sufficient to prevent severe complications due to anorexia in the short term.

DOI: 10.1201/9781003308270-38

EMERGENCY CARE BY A VETERINARY PROFESSIONAL

Anorectic birds are stabilized using nutrition, fluid therapy, heat and (in case of dyspnea) oxygen (see General stabilization of sick birds, p. 23). Intake of energy and fluids is essential in this situation and assisted (forced) feeding should be performed immediately when birds no longer produce feces and do not eat at free will.

SYMPTOMATIC TREATMENT OF VOMITING

In case of nausea, metoclopramide or maropitant (parenteral) is indicated prior to the possible assisted feeding.

Loss of appetite and anorexia are non-specific symptoms of the disease. Therefore, after stabilization, a complete physical examination should be performed to diagnose the underlying cause of the significant decrease in the amount of feces. Further treatment depends on the cause.

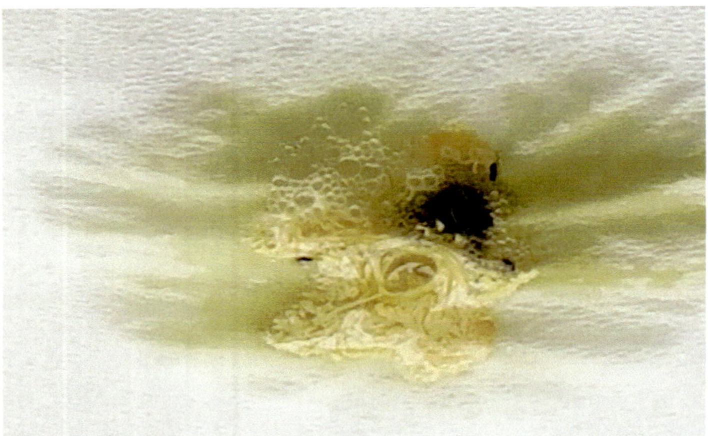

Fig. 35.1 Droppings with decreased amount of feces.

BLACK FECES

Black discoloration of feces (**Fig. 35.2**) can be caused by the presence of digested blood. This is normal in birds that eat meat. In other species, black discoloration of the feces can be a sign of blood loss in the oral cavity, throat, oesophagus, crop, proventriculus (glandular stomach), ventriculus (gizzard) or small intestine. Bleeding in the oral cavity, oesophagus, throat or crop usually results in vomiting or bloody discharge from the beak also, so when these symptoms are absent, black discoloration of feces is most likely caused by bleeding in the proventriculus, ventriculus or small intestine. In addition to causes like inflammation of the small intestine or stomach, tumors, foreign bodies, intoxication and liver disease, black discoloration of feces can also be due to anorexia—especially in small birds, where lack of food in the intestines for more than 24 hours can lead to spontaneous bleeding in the intestines (hemorrhagic diathesis). In this situation, decreased fecal volumes will have preceded the black discoloration. Unless the blood loss originates in the oral cavity or is clearly caused by anorexia, diagnosing the cause of the gastrointestinal bleeding can be challenging.

ADVICE FOR FIRST AID AT HOME

When black feces is the result of anorexia, owners should offer favorite food items and try to stimulate/encourage the bird to eat. Intake of energy is most important and, initially, for a short term, the exact nutrient composition of the food is of minor importance as long as the food items being offered are not harmful. For example, psittacines can be offered sunflower seeds or spray millet. When tame birds don't eat at all but are well awake, liquid food may be supplied with a syringe or spoon into the beak on the middle of the tongue (not all the way back in the throat, as this can lead to aspiration and death). In general, 2 ml of liquid diet per 100 grams of body weight is sufficient to prevent severe complications due to anorexia in the short term.

Fig. 35.2 Black feces indicates blood loss in the gastrointestinal tract.

EMERGENCY CARE BY A VETERINARY PROFESSIONAL

General stabilization

Birds with black feces must be stabilized using fluid therapy, heat and nutrition (see General stabilization of sick birds, p. 23). Anorexic birds with black discoloration of the feces should be fed immediately by means of assisted (forced) feeding.

Diagnostics

Unless absence of food has been identified as the cause of black feces, a complete physical examination should be performed after stabilization in an attempt to diagnose the cause of the black feces.

X-rays

X-rays (see Appendix 3, p. 179) are indicated to check for the presence of radiodense foreign bodies, space-occupying lesions or metal particles (see Lead poisoning, p. 73).

Fecal examination

Fecal samples are taken for microscopic examination (see Appendix 4, p. 191).

Blood tests

Ideally, blood should be collected (see Observation, physical examination and diagnostic tests, p. 13 and Appendix 2, p. 169) for hematology, biochemistry and toxicology.

Note: In very small species or seriously ill birds, collecting blood is sometimes too dangerous, especially by veterinarians less trained in collecting blood from birds. Too much blood loss, formation of hematomas, handling for too long or excessive stress can lead to death in small or unstable patients. In some cases, blood collection is best not done during the emergency consultation.

THERAPY

Protectant

Sucralfate (applied 2 hours apart from other medication or food) can help stop gastrointestinal bleeding by forming a coating over gastrointestinal ulcers.

Symptomatic treatment of vomiting

In case of vomiting, anti-emetics (e.g., maropitant) are indicated. Metoclopramide is contraindicated in case of gastrointestinal bleeding.

Antibiotics

Antibiotic treatment (e.g., amoxicillin/clavulanic acid or TMPS) can be considered as long as no other non-bacterial cause has been determined. It can be curative in case of bacterial infections and prevent secondary bacterial overgrowth in case of other causes of black feces.

Bigger birds can be referred to an avian specialist for gastroscopy.

YELLOW OR GREEN URATES

Yellow or green discoloration of urates in fresh droppings (**Fig. 35.3**) may indicate liver disease. Liver disease may be caused for example by infections (including psittacosis), hepatolipidosis (fatty liver disease) or tumors. Urates in older droppings can be colored due to contact with feces.

EMERGENCY CARE BY A VETERINARY PROFESSIONAL

Note: Yellow or green urates can be a sign of the zoonotic disease psittacosis (see Appendix 11, p. 225).

DIAGNOSTICS

Diagnosing the cause of yellow or green urates can be challenging. X-rays (see Appendix 3, p. 179), blood tests (see Observation, physical examination and diagnostic tests, p. 13), ultrasound and liver biopsies may all be necessary, but in most cases will not be possible during an emergency consultation. Samples taken with dry swabs from the choana and cloaca can be submitted for polymerase chain reaction (PCR) testing for *Chlamydia psittaci* (psittacosis).

Ideally, blood should be collected (see Observation, physical examination and diagnostic tests, p. 13 and Appendix 2, p. 169) for hematology and biochemistry.

Note: In very small species or seriously ill birds, collecting blood is sometimes too dangerous, especially by veterinarians less trained in collecting blood from birds. Too much blood loss, formation of hematomas, handling for too long or excessive stress can lead to death in small or unstable patients. In some cases, blood collection is best not done during the emergency consultation.

TREATMENT

General stabilization

Birds with yellow or green urates are usually severely ill and should be stabilized using heat, fluids and nutrition (see General stabilization of sick birds, p. 23).

Antibiotics

Treatment with antibiotics is indicated in cases of bacterial infections or psittacosis and could be considered in birds with yellow or green urates from unknown causes. In case of suspicion of psittacosis, doxycycline is the first choice of antibiotic and therapy with doxycycline should be started pending results of the tests.

Note: Intramuscular injections of doxycycline cause severe damage at the injection site and should only be considered in situations where oral administration is not possible or ineffective (for example in case of vomiting, crop stasis, in parents feeding crop contents to juveniles or when birds cannot be caught and held).

When there is no suspicion of psittacosis, amoxicillin/clavulanic acid can be a better option, as doxycycline is potentially hepatotoxic, especially in a case of already decreased liver function.

Note: Blood chemistry is an important diagnostic tool for diagnosing the cause of yellow or green urates. Aspartate aminotransferase (AST) is an important marker of liver cell damage. Unfortunately, AST is not only located in liver cells, but also in muscle cells.

Damage to muscles caused by intramuscular injections will cause an elevation in AST in the blood. When blood collection is postponed until after the initial emergency treatment, not giving intramuscular injections will make interpretation of blood chemistry in later stages more reliable. Of course, improving changes of survival is most important, and injections should be given when necessary to stabilize emergency patients.

Fig. 35.3 Droppings with yellow urates on the right.

DIARRHEA

Diarrhea (very liquid feces, **Fig. 35.4**) should be distinguished from polyuria (highly watery urine) in birds.

Diarrhea can have multiple causes, including infections (with bacteria, parasites, viruses, yeasts and Chlamydia), intoxications and liver disease. Diarrhea can lead to malnutrition, dehydration and hypothermia by loss of water, electrolytes and nutrients.

ADVICE FOR FIRST AID AT HOME

Birds with diarrhea lose lots of fluids. Birds should be offered clean drinking water at all times to attempt to compensate for fluid loss. Drinking more can prevent or minimize dehydration. Owners should never withhold food in an attempt to stop diarrhea, as this can lead to dangerous situations.

EMERGENCY CARE BY A VETERINARY PROFESSIONAL

Note: Diarrhea can be a sign of the zoonotic disease psittacosis (see Appendix 11, p. 225).

FECAL EXAMINATION

Fecal samples are taken for microscopic examination (see Appendix 4).

If complete fecal examination during the emergency consultation is not possible, fecal samples and smears are kept for later examination.

Overgrowth of just one type of bacteria (monoculture) and, for example, overgrowth of Clostridia (bacteria forming endospores and sometimes resembling miniature tennis rackets) can be relevant in case of abnormal droppings. Unfortunately, not all pathogenic bacteria can be identified just by microscopic examination or will form monocultures. On the other hand, abnormal bacterial flora is not always clinically relevant and can be secondary to underlying pathology. Fecal samples can be sent to a laboratory for bacteriological culturing and susceptibility testing in case of suspicion of bacterial infection.

Note: Birds of most species do have bacterial intestinal flora, so the presence of bacteria in the feces of most birds is normal.

Parasitic infections are quite common in free ranging birds (in many cases not clinically relevant) and outdoor aviary birds (frequently clinically relevant), but uncommon in birds living indoors (usually clinically relevant).

Overgrowth of yeasts can be clinically relevant, but is in most cases secondary to underlying pathology or a compromised immune system.

RADIOLOGY

X-rays (see Appendix 3, p. 179) or computed tomography (CT scan) can be used to check for the presence of metal particles in the gastrointestinal tract (see Lead poisoning, p. 73), enlarged organs, abnormal gastrointestinal motility or masses.

TREATMENT

General stabilization

Birds with diarrhea are stabilized using nutrition, fluids and heat (see General stabilization of sick birds, p. 23). At least half of the fluids should be given parenterally, because of the possibly less functional gastrointestinal tract. The other half can be administered into the crop in the form of Oral Rehydration Solution (ORS) or, for example, NaCl/2% glucose through a feeding tube (see Appendix 5, p. 195). In cases of crop stasis, all fluids should be given parenterally.

Treatment of infection or overgrowth of microorganisms

Flagellates can be treated with oral anti-protozoal drugs from the nitroimidazoles class (e.g., metronidazole, ronidazole). Worm infections can be treated with anthelmintics (e.g., fenbendazole, flubendazole, ivermectin, praziquantel). Coccidia can be treated with anti-coccidial drugs (e.g., toltrazuril, sulfadimethoxine).

Overgrowth of yeasts is treated with anti-fungal agents. Nystatin and amphotericin B are generally effective and relatively safe to use because of the minimal systemic absorption after oral intake. Itraconazole can be used as well, but is systemically absorbed and can cause more side effects.

Bacterial infections/overgrowth of bacteria are treated with antibiotics (e.g., TMPS or amoxicillin/clavulanic acid). Antibiotic treatment is started pending culture results.

Note: Antibiotic treatment (e.g., amoxicillin/clavulanic acid or TMPS) can be considered during emergency consultations in any case of diarrhea as long as no other non-bacterial cause has been determined. It can be curative in cases of some bacterial infections and prevent secondary bacterial overgrowth in cases of other causes of diarrhea.

Further diagnostics and treatment are indicated when the diarrhea continues.

Fig. 35.4 Diarrhea.

FRESH BLOOD

Fresh blood in droppings (**Fig. 35.5**) is caused by blood loss in the last part of the gastrointestinal tract or urogenital system. Causes include infections, cloacal prolapse (see p. 85), egg binding (see p. 105), tumors, papillomas, cloacal trauma, lead poisoning (see p. 73), blood clotting diseases and cloacoliths (accumulation of dried urates in the cloaca).

Fig. 35.5 Droppings with fresh blood.

EMERGENCY CARE BY A VETERINARY PROFESSIONAL

Fecal examination

Fecal samples are taken for microscopic examination (see Appendix 4, p. 191 and Diarrhea, p. 150).

If complete fecal examination during the emergency consultation is not possible, fecal samples and smears are kept for later examination.

Fecal samples can be sent to a laboratory for bacteriological culturing and susceptibility testing in case of suspicion of bacterial infection.

External inspection of the vent

The vent is checked for cloacal prolapse (see p. 85) and external injuries.

Palpation

The coelom/abdomen is palpated and a cloacal toucher is performed with a finger (in large birds like chickens) or cotton tip applicator to check for masses, cloacoliths and eggs.

X-rays

X-rays (see Appendix 3, p. 179) are indicated to check for the presence of calcified eggs (see Egg binding, p. 105), cloacoliths and for the presence of metal particles in the gastrointestinal tract (see Lead poisoning, p. 73).

Internal inspection

Under general anesthesia (e.g., with isoflurane/sevoflurane), the internal parts of the cloaca can be checked for trauma, intussusception, papillomas, cloacoliths and tumors.

TREATMENT

General stabilization

The health status of birds with fresh blood in their droppings varies with the amount of blood loss and underlying pathology. Sick birds must be stabilized using fluid therapy, heat and nutrition (see General stabilization of sick birds, p. 23).

Antibiotics and anti-parasitic treatment

In case of injuries to the cloaca or bacterial overgrowth, antibiotic treatment (e.g., amoxicillin/clavulanic acid) is indicated.

Note: Antibiotic treatment can be considered during emergency consultations in any case of bloody droppings as long as no other non-bacterial cause has been determined. It can be curative in cases of some bacterial infections and prevent secondary bacterial overgrowth in cases of other causes of bloody droppings.

Parasitic infections should be treated. Flagellates can be treated with oral anti-protozoal drugs from the nitroimidazoles class (e.g., metronidazole, ronidazole). Worm infections can be treated with anthelmintics (e.g., fenbendazole, flubendazole, ivermectin, praziquantel). Coccidia can be treated with anti-coccidial drugs (e.g., toltrazuril, sulfadimethoxine).

Analgesia

NSAIDs (e.g., meloxicam) are indicated in case of injuries and tenesmus. NSAIDs can be contraindicated in other causes of bloody droppings, so do not start treatment with NSAIDs when the cause has not been determined.

Protectant

Sucralfate (applied 2 hours apart from other medication or food) can help stop gastrointestinal bleeding by forming a coating over gastrointestinal ulcers.

Surgery

Deep lesions of the cloaca can be closed with monofilament sutures (usually 4–0 to 5–0) under general anesthesia (e.g., with isoflurane/sevoflurane) and adequate analgesia (e.g., local lidocaine and meloxicam). When traumatic lesions affect the cloacal mucosa and the skin, the mucocutaneous junction is aligned first.

Cloacoliths can sometimes be removed as a whole or after crumbling them to smaller pieces with pliers through the cloaca in patients under general anesthesia. In other cases, surgery is necessary for safe removal of cloacoliths. Patients should be stabilized before (referring them for) surgery.

Diffuse bleeding surfaces of papillomas or tumors can be touched with a silver nitrate stick in an effort to temporarily stop the bleeding. Cloacal surgery for removing masses is challenging and is best done by an avian specialist.

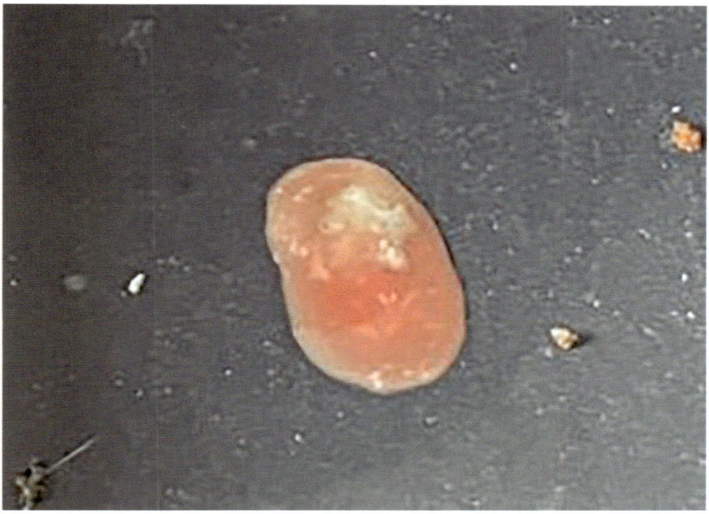

Fig. 35.6 Pink urates/urine.

PINK URATES

A pink discoloration of the urates (**Fig. 35.6**) can be caused by severe kidney damage and/or lead intoxication.

EMERGENCY CARE BY A VETERINARY PROFESSIONAL

Diagnostics

Blood test

Ideally, blood is collected (see Observation, physical examination and diagnostic tests, p. 13 and Appendix 2, p. 169 for accessible veins and technique) for hematology, blood lead levels and uric acid levels.

Note: In very small species or seriously ill birds, collecting blood is sometimes too dangerous, especially by veterinarians less trained in collecting blood from birds. Too much blood loss, formation of hematomas, handling for too long or excessive stress can lead to death in small or unstable patients. In some cases, blood collection is best not done during the emergency consultation.

X-rays

X-rays (see Appendix 3, p. 179) are indicated to check for the presence of metal particles in the gastrointestinal tract (see Lead poisoning, p. 73).

Note: Absence of visible metal particles on X-rays does not rule out lead poisoning completely.

Treatment

In cases of proven lead intoxication or a strong suspicion of lead intoxication based on the medical history, symptoms, and/or X-ray findings, treatment must be started immediately, see Lead poisoning, p. 73.

Fluid therapy

Fluid therapy (see p. 27) is indicated in cases of lead intoxication and renal disease. Since lead intoxication can lead to a less functional gastrointestinal tract, fluids should be administered parenterally.

Antibiotics

When no metal particles are visible on X-rays and blood lead levels have not yet been determined and/or hematology indicates an inflammatory reaction, treatment with non-nephrotoxic broad-spectrum antibiotics (e.g., amoxicillin/clavulanic acid) could be considered, as a bacterial nephritis is another possible cause of kidney damage.

36 *Damaged air sac*

DAMAGE to the wall of air sacs, caused by trauma or chronic infection, leads to air leakage from the respiratory system to the subcutaneous tissues (emphysema). This is most common in the neck region, but damage to the body wall can lead to leakage on other locations. In cases of rupture of the air sac in the neck, the neck will be distended (**Fig. 36.1**). In cases of generalized emphysema, other parts of the bird might appear swollen.

Emphysema leads to discomfort, pain and, in severe cases, dyspnea and anorexia.

Fig. 36.1 Pigeon with hyperinflated cervicocephalic air sac and subcutaneous emphysema.

DOI: 10.1201/9781003308270-39

EMERGENCY CARE BY A VETERINARY PROFESSIONAL

Swelling of the affected body part can be observed. Through palpation, soft swellings can be felt and sometimes also crackling sensations. Air can be aspired from the swelling.

Note: Hyperinflation of the cervicocephalic air sac and distension of the crop caused by swallowing air (aerophagia, most common in young chicks) can also cause local, air filled swellings in the neck.

For relief, the skin and (in case of hyperinflation of the cervicocephalic air sac) air sac wall can be perforated. Under general anesthesia (e.g., isoflurane/sevoflurane), a transparent and avascular part of the skin of the bloated area is disinfected. The skin can be punctured with a needle or scalpel to let the air out (**Fig. 36.2**). Recurrence is common after a single deflation. Removing a part of the skin and air sac wall (fenestration) will lower the chance of immediate recurrence. Further diagnostics and therapy are needed after stabilization of the patient, as underlying pathology is likely.

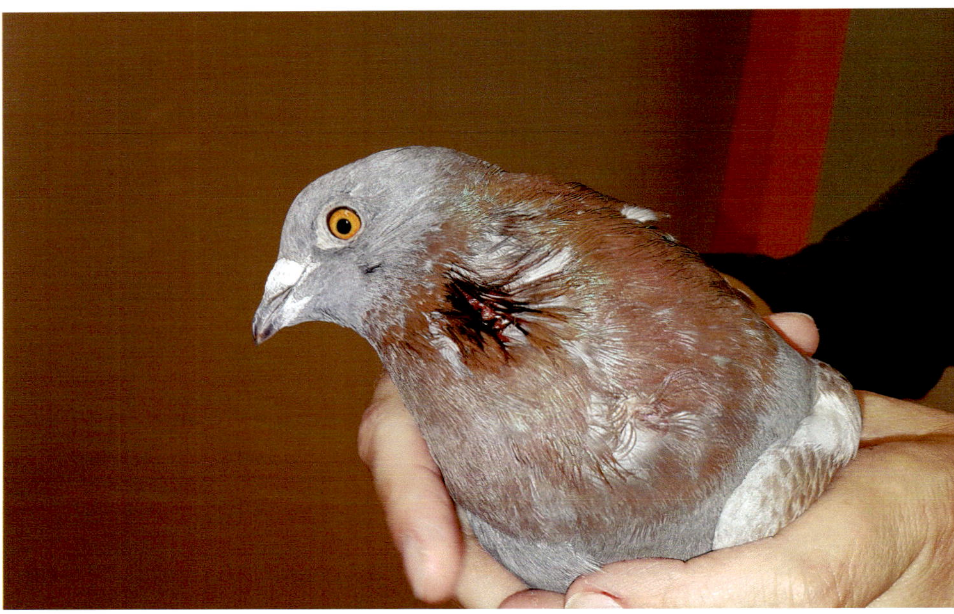

Fig. 36.2 The same pigeon after fenestration of the skin and air sac wall.

Appendices

A1 – Technique: Handling of birds

Birds are fragile and easily stressed. They should be handled in an efficient way for examination and treatment without causing harm to the patient or staff and without unnecessary stress.

With the exception of birds of prey, owls, pigeons, poultry, waterfowl and other large non-psittacine species, most birds are best held for examination and treatment by securing the head and wrapping the rest of the body in a (paper) towel (**Figs. A1.1 and A1.2**). The patient's body should be loosely wrapped in a towel, so breathing is not restricted, while the wings and legs are secured.

The person handling the bird holds it close to their body in an upright position with the head up. The head is held with light pressure between the thumb and index or middle finger on either side of the head/lower jaw. When using the thumb and middle finger for this, the index finger can be placed on top of the head for extra control (**Fig. A1.3**). The bird is positioned with its back against the palm of the hand. The other hand is held around the wrapped belly, lower back and wings. Very small birds are handled in about the same way, with the difference that they are held with just one hand with the ring finger and little finger kept loosely around the lower body instead of the other hand.

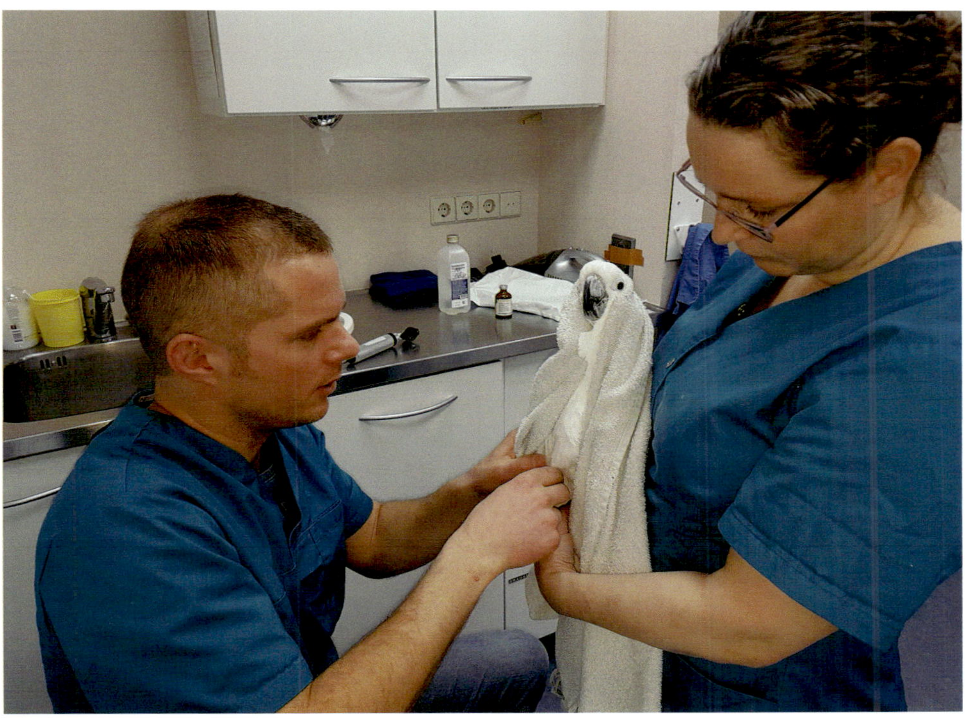

Fig. A1.1 Cockatoo being held for physical examination.

Fig. A1.2 Most parrots are best held while wrapped in a towel with light pressure on both sides of the head/lower jaw.

PSITTACINES: PARROTS AND PARAKEETS

Handling psittacines can be challenging, as they are strong, smart and able to cause considerable damage—and at the very least pain—by biting. Most tame parrots and parakeets are not aggressive, but will bite in self-defense. A calm and respectful approach usually makes handling easier and less stressful.

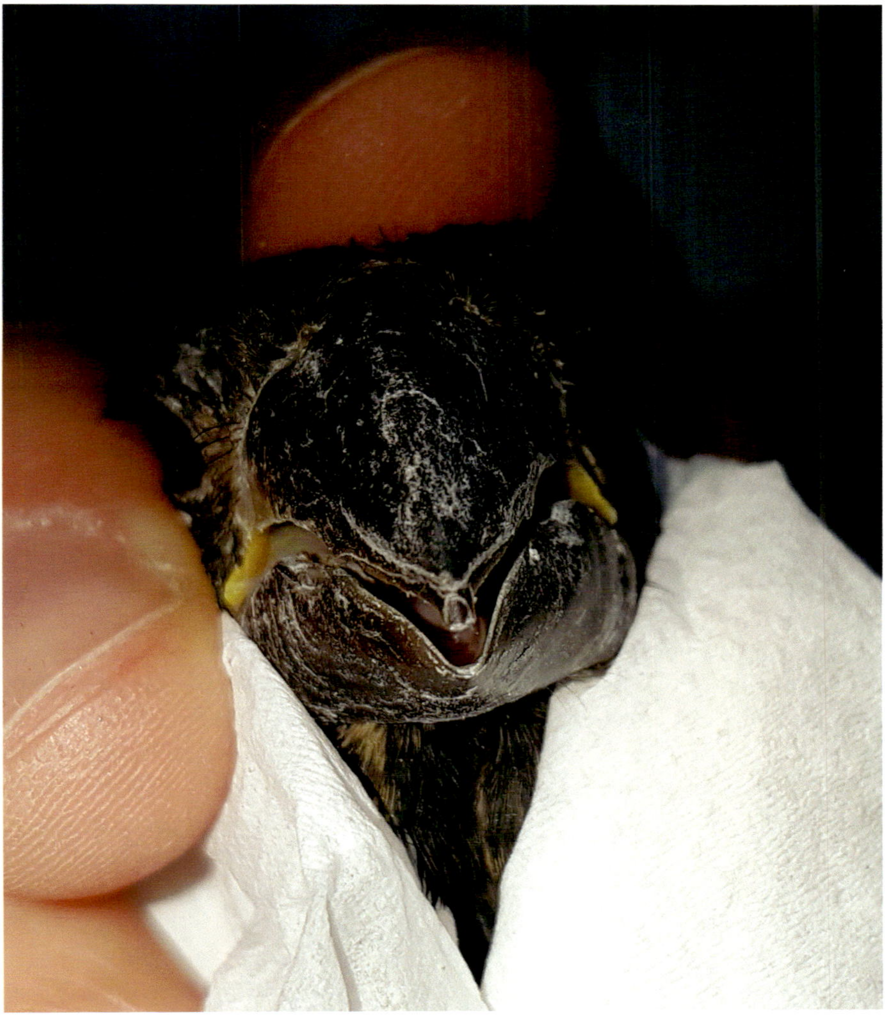

Fig. A1.3 Small birds can be held with a paper towel. In this photo of the great-billed seed finch, the index finger is placed on top of the head for extra control.

There are several methods for successfully getting psittacines in the correct position in the towel.

Very tame birds (that do not tend to fly away) can best be taken out of the cage or carrier by the owner. The owner can hold the bird close to his/her body and prevent it from flying away by laying a hand over the shoulders/wings. Then, someone else can gently take the bird with a towel. First, the head is secured, then the towel is wrapped around the body and the bird is lifted.

Another way to get tame birds (that do not tend to fly away) in the towel is to let them step up from the cage or transport box onto a hand, arm or stick and subsequently step off onto the table. People present can stand closely around the table to prevent the bird from falling off the table or flying away. Next, a towel is placed over the bird's entire body. This is done gently but in a purposeful way. The head is then secured through the towel, followed by wrapping the towel around the body and lifting the bird.

A third technique to get tame larger psittacines into the towel is explained below. When used calmly, this technique is less scary for many parrots than being taken out of the cage or up from the table with a towel. Be careful when using this technique, as it does take some practice and can lead to fractures of the lower legs if not performed correctly. This technique is therefore not advised for people with less experience with birds.

A towel is held at one corner between the ring finger and little finger of the right hand in a right-handed handler. The index finger is offered to be stepped on by the parrot. When the parrot is sitting on the index finger, the handler holds the cranial toes by laying the thumb over the toes. The bird is kept close to the body of the handler to prevent it from falling over or trying to escape. The towel is lifted from behind with the other hand and gently draped over the bird. The head is secured through the towel and the towel is wrapped around the rest of the body.

Birds that do not step up voluntarily, tend to fly away (the majority of the smaller species) and birds that are very excited, scared or aggressive may best be taken out of the cage or transport box directly by the person who is going to handle them. Objects that might hinder getting the bird (e.g., food and water bowls, toys, hiding places and perches) should first be removed. Dimming the light can also help. A (paper) towel is draped over the open hand to prevent the bird from seeing the individual fingers and thereby preventing aimed bites. Put the hand with the towel over the bird's shoulders and neck and secure the head through the towel. At the same time, control the rest of the body—especially the wings—with the other hand using the towel to prevent the bird from making wild movements while only being held at the head and neck.

BIRDS OF PREY AND OWLS

Birds of prey are heavily armed. While in large species damage can be caused by the beak (**Fig. A1.4**), the most danger comes from the talons/claws (**Fig. A1.5**). When handling birds of prey and owls, the legs must be secured at all times to prevent serious injury to the staff.

Fig. A1.4 Birds of prey and owls, especially the bigger species, can cause injury by biting as well.

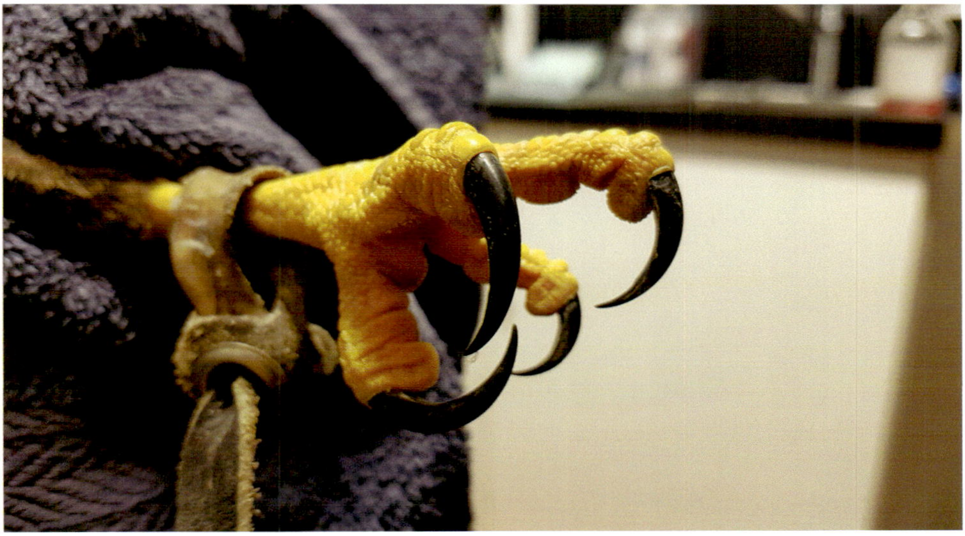

Fig. A1.5 Birds of prey and owls can cause severe injuries with their talons. The legs should be secured at all times.

Most human-kept larger birds of prey and owls can be taken out of the transport box by the owner. By keeping the jess (leash) attached to the anklets pulled short, birds can be held on the hand protected by a falconry glove. While the owner controls the feet at the glove, a towel can be used to cover and control the wings, body and head of the birds. Covering the eyes can help calm down stressed patients. Next, the legs are secured, preferably one leg per hand (**Fig. A1.6**), but if necessary, both legs can be held with one hand. In that case, a finger is placed between the legs to prevent damage caused by pushing the legs hard onto each other. Large species are best held with their back resting against the body of the handler with both legs being held and their wings and head covered with a towel.

Fig. A1.6 Both legs are held at the tarsometatarsus separately.

During anesthesia, the legs should still be held or the talons must be wrapped in elastic self-adhesive bandage to prevent injuries to the personnel if the bird wakes up unexpectedly.

PIGEONS

Pigeons can best be held facing the handler. The legs are held between the index and middle finger and the lower back and wing tips are controlled by the thumb (**Fig. A1.7**).

Fig. A1.7 Handling of a pigeon.

POULTRY, WATERFOWL AND MOST OTHER BIG BIRDS

Most birds from these categories can generally be examined and treated while standing on the table. A towel can be placed over the wings to prevent flapping if necessary. Meat-type chickens and ducks are especially prone to stress-induced hyperthermia.

Note: Almost every bird can bite or peck during handling. Not only psittacine birds, birds of prey and owls can cause damage with their beaks. For example, seagulls, coots, herons and cormorants can cause injuries and are, in general, not the most cooperative patients. Keep them at a distance from the face and protect the hands and eyes.

A2 – Technique: Subcutaneous, intravenous and intraosseous infusion and venipuncture

SUBCUTANEOUS INFUSION

Fluids can be administered as a bolus by subcutaneous injection. There are three sites where subcutaneous infusion can be administered in birds: The left and right inguinal fold and the interscapular region (on the back between the shoulder blades). The administration of fluids into the inguinal fold appears to be less painful for the birds. However, it can be a bit more technically demanding and the patient must be well restrained to minimize the risks of entering the abdominal cavity or thoracic air sac with the needle. Subcutaneous administration in the interscapular region appears to be more painful, but is easier to administer. With some practice, this route may be feasible in many birds without assistance.

TECHNIQUE FOR SUBCUTANEOUS INJECTION IN THE INGUINAL FOLD

The patient is restrained with the abdomen/chest facing the person who will be administering the fluids. The leg on the side of the injection is extended caudally. In this position, after moistening the feathers and skin with a small amount of alcohol (to minimize cooling), the inguinal fold is visible between the thigh and the body wall. With a thin needle (for example 25G/orange needle) fluids can then be injected between the two skin layers (**Fig. A2.1**). Care should be taken not to perforate the abdominal wall with the needle. Injecting with the needle pointed away from the bird's coelom/abdomen and placing a sterile finger on the needle so that a deep injection is impossible will minimize this risk. To avoid excessive tension on the skin and discomfort caused by administration of too large a volume, the fluid can be distributed over both inguinal folds.

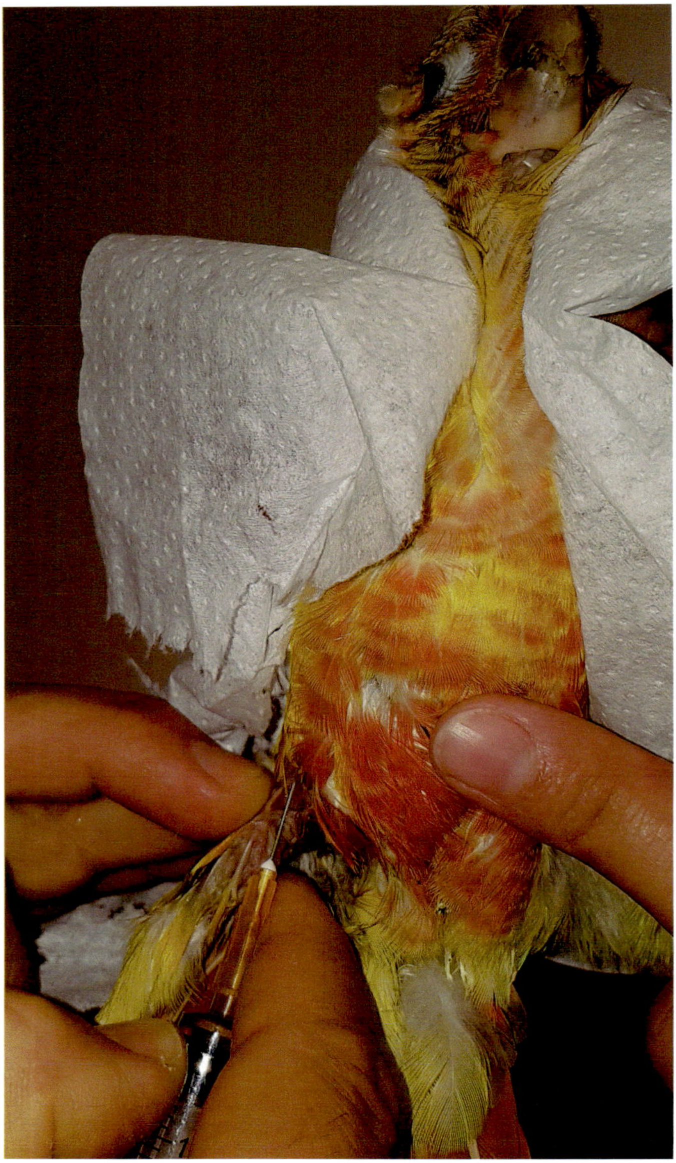

Fig. A2.1 Subcutaneous injection in the inguinal fold of a pyrrhura.

TECHNIQUE OF SUBCUTANEOUS INJECTION IN THE INTERSCAPULAR REGION

The patient stands on the table in a natural position or is lying down in sternal recumbency. The bird is held in place with a (paper) towel. While—especially—the head and lower back and wings are restrained, the (paper) towel is then folded forward, exposing the back and shoulder region. Fluids are administered in the interscapular region. The neck is extended to prevent injection in the neck area instead of the interscapular region

Note: Because of the presence of air sacs in the neck, injecting fluids cranial to the interscapular region could lead to drowning by administering the fluids into the respiratory system and should be avoided.

The wings and lower back are now held with a bare hand. The cover feathers in the area between the shoulder blades are moistened with a small amount of alcohol (to minimize cooling) and then pushed aside.

The skin of birds in this region is thin and there is hardly any subcutaneous tissue. An injection through the skin quickly results in damage to deeper structures and unnecessary pain if given too deeply. To prevent this, a small amount of fluid can be injected through the needle before and during the perforation of the skin. Once the tip of the needle has passed the skin, this will create a subcutaneous bubble of fluid (**Fig. A2.2**). Keeping the tip of the needle in this bubble while depressing the syringe further prevents damage to the deeper tissues. When perforating the skin, ensure the bevel of the needle is observed. Most small to medium sized birds can be injected well with a short 25G/orange needle.

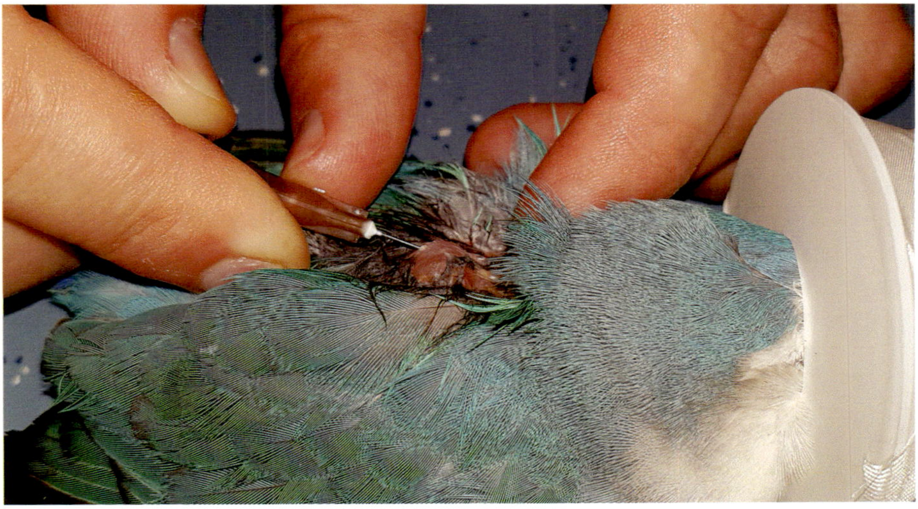

Fig. A2.2 Subcutaneous injection in the interscapular region of a sedated lovebird.

VENOUS ACCESS

Venous access allows sampling of blood as well as administration of fluids.

There are three veins that can best be used in birds for blood sampling or intravenous infusion (and venous blood sampling):

- Right jugular vein.
- *Basilic vein*: Passes the elbow at the bottom of the wing.
- *Medial metatarsal vein*: Located on the medial side of the lower leg.

Not every vein is equally useful in every bird species. Only use a vein that is clearly visible and in which the catheter is easily stabilized.

In some cases, such as in severely stressed or agitated patients or when the procedure is done by a less-experienced veterinarian, giving the intravenous injection or sampling blood from the veins is safest while the bird is sedated. Sedation can be accomplished by the use of e.g., isoflurane/sevoflurane or midazolam combined with butorphanol.

VENOUS ACCESS TO THE RIGHT JUGULAR VEIN

For venipuncture of the jugular vein, the bird is standing in a normal position on the table or lying down in sternal recumbency. In birds that are not sedated, an assistant will hold the body of the bird with a towel or piece of paper, while the person performing the venipuncture restrains the head with the hand. A right-handed person restrains the head and neck of the bird with the left hand to inject in the vein on the right side of the neck. The left jugular vein is a much smaller blood vessel and therefore less suitable. Fortunately, the skin on the side of the neck is bare in most bird species. This is not the case with water birds, however. The cover feathers that lie over this bald spot are moistened with a small amount of alcohol and then pushed to the side, so that the skin becomes visible. The skin itself is so thin that in most (non-obese) birds, the deeper structures are clearly visible. The right jugular vein is a large vein that is relatively easy to locate visually. The head and neck must be kept in the correct position for this. The neck is held with the index finger and thumb on the upper side and the middle finger on the lower side. The index finger pushes the head a little downward to the left (the neck is bent a bit with left nostril pointing to the table), the middle finger pushes the jugular vein up from below to the surface (the middle finger therefore pushes up from below on the right side of the cervical vertebrae) and with the thumb, light pressure is put on the vein to make it enlarge with blood, ready for venipuncture (**Fig. A2.3**).

Fig. A2.3 Right jugular vein.

VENOUS ACCESS TO THE BASILIC VEIN

The basilic vein is a relatively large blood vessel on the underside of the wing. At the elbow, the vein crosses from the lower wing to the upper wing and can be followed to the armpit. The cover feathers are moistened with a small amount of alcohol and then pushed aside to reveal the skin and deeper structures and to ensure disinfection. Although the vein can usually also be visualized in a bird that is simply standing upright with the wing raised, this position often results in a decreased filling of this blood vessel. Usually, the administration of infusion fluid into the basilic vein is easier in a bird lying on its back (**Fig. A2.4**). This requires either manual restraint or sedation. After making the vein enlarge with blood by putting light pressure on the vein proximal to the puncture site, the blood vessel can be punctured relatively easily in a non-moving bird. A disadvantage of using the basilic vein for administering fluids or collecting blood is that hematomas may occur quickly, either by damage to the vessel wall due to movement of the needle relative to the patient or by bleeding after removal of the needle or catheter. In birds with normal blood clotting, it is often necessary to keep pressure on the injection site for a few minutes. Hematomas at this location can become so large that the intravascular volume gain from the infusion can be significantly decreased by the subcutaneous bleeding.

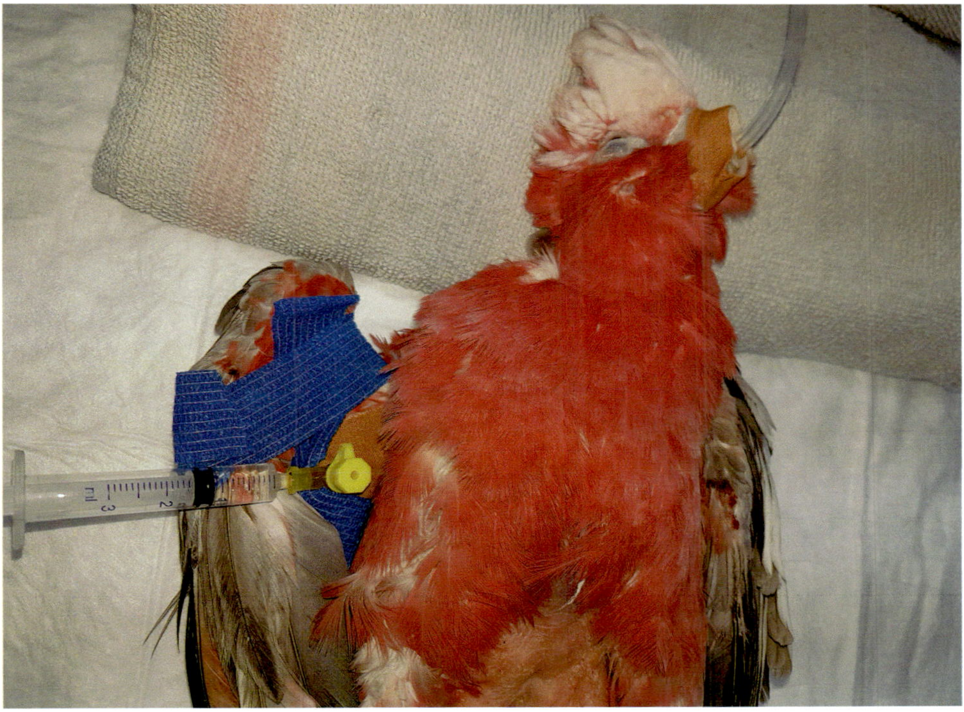

Fig. A2.4 IV infusion in the basilic vein.

VENOUS ACCESS TO THE MEDIAL METATARSAL VEIN

The medial metatarsal vein is located on the medial side of the tarsometatarsus. The vein is particularly suitable for intravenous infusion and blood collection in somewhat larger birds with a long tarsometatarsal bones, such as many waterfowl, some birds of prey and gallinaceous birds (such as chickens). Fixation of a catheter on a leg is easier than on a wing, so that a continuous infusion using an infusion pump is often possible when using the medial metatarsal vein. Unfortunately, this vein is often not very useful for the placement of a catheter in psittacines and small species.

The patient is restrained in a lateral position. The leg close to the table is being used for venipuncture, as the medial metatarsal vein is now on top and visible. The skin is cleaned if it is dirty and then disinfected with alcohol. By light pressure above the hock joint (tibiotarso-tarsometatarsal joint), the vein is enlarged with blood and therefore more visible. An intravenous catheter (26G, 24G or 22G) can be used for longer infusion, or a hypodermic needle (for example a 25G needle) can be used to give a bolus (**Fig. A2.5**). When a bird has hard dermal scales covering the leg, the needle or catheter should be inserted through the skin between the scales.

Fig. A2.5 Injection in medial metatarsal vein of a chicken.

INTRAOSSEOUS INFUSION

Due to the painful nature of the procedure, insertion of an intraosseous catheter without analgesia is seriously detrimental to the patient's well-being. This procedure is therefore preferably done under general anesthesia with systemic analgesia (e.g., butorphanol) and local analgesia (e.g., with lidocaine or bupivacaine). Sometimes, due to an acutely life-threatening situation, there is no appropriate option for complete analgesia with systemic agents. In such a case, anesthesia of the soft tissues and periosteum at the site of the infusion will have to be provided at least by means of a local analgesic agent (e.g., lidocaine or bupivacaine).

TECHNIQUE FOR INTRAOSSEOUS CATHETER PLACEMENT IN THE SHIN BONE (TIBIOTARSUS)

The knee region is aseptically prepared after plucking a few feathers covering the incision site. The use of large amounts of alcohol during disinfection is avoided to prevent (further) hypothermia. After bending the knee, a small puncture incision is made through the skin just medial to the patellar tendon. Through this incision, the tip of a hypodermic needle or spinal needle with stylet (this is ideal, because bone tissue can obstruct the lumen of a normal hypodermic needle) is placed on the tibial plateau. The needle is then inserted into the medullary cavity with a twisting motion in the longitudinal direction of the bone.

TECHNIQUE FOR INTRAOSSEOUS CATHETER PLACEMENT IN THE ULNA

The wrist region (**Fig. A2.6**) is aseptically prepared after extracting a few feathers covering the incision site (**Fig. A2.7**). After flexion of the wrist, the crest of the distal ulna is located by palpation. After making a small puncture incision through the skin, the tip of a hypodermic needle or spinal needle with stylet is placed in longitudinal direction on the distal surface of the ulna. While the ulna is being fixated, the needle is inserted with a twisting motion into the medullary cavity of the ulna (**Fig. A2.8**). Since the ulna is also connected to the respiration system in a small number of species (for example some vultures and storks), it is important to aspirate first. When air is aspirated, the administration of fluid therapy in the ulna is contraindicated. X-rays can be used to check correct placement (**Fig. A2.9**).

Fig. A2.6 Dorsal side of the distal wing of an amazon parrot.

Fig. A2.7 The area is prepared for the intraosseous infusion by plucking the feathers, administering a local analgesic and disinfecting the skin.

Fig. A2.8 Needle inserted in the ulna.

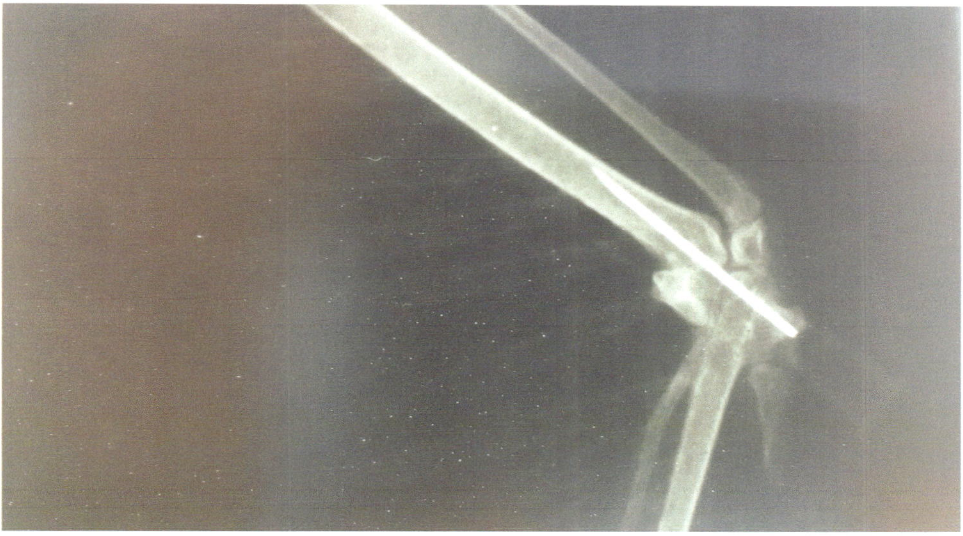

Fig. A2.9 X-ray showing the needle placed correctly in the ulna.

A3 – X-rays

X-rays should always be performed with sufficient protection for the personnel against X-ray radiation. Holding patients while performing X-rays is not acceptable without wearing X-ray gloves. Handling birds with protective X-ray gloves is, in general, not possible without risks of escape, trauma, extreme stress and suboptimal positioning (**Fig. A3.1**).

As a result of this, well-positioned X-rays of birds are in almost all cases best made under anesthesia (e.g., using isoflurane/sevoflurane or midazolam combined with butorphanol).

Ideally, both correctly positioned laterolateral and ventrodorsal full body X-rays are made.

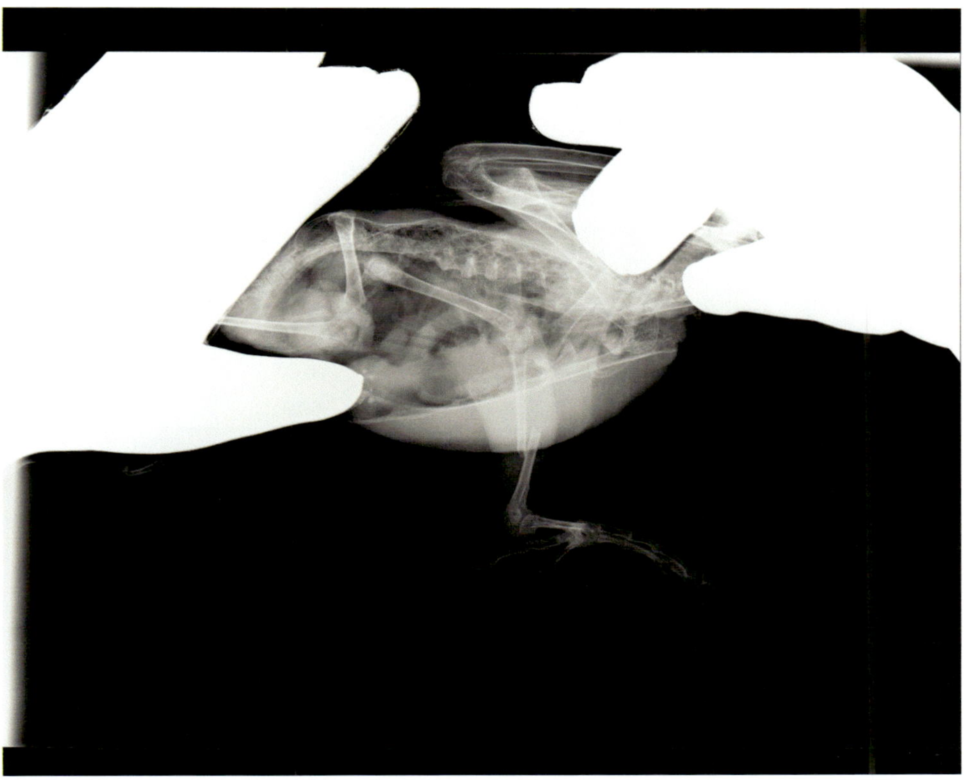

Fig. A3.1 Example of incorrect technique for taking X-rays of a bird. Lead gloves are not suited for holding birds. Positioning is incorrect and the hands of the practitioner are in the primary X-ray beam (lead gloves do not provide enough protection against the primary beam).

LATEROLATERAL PROJECTION

For the laterolateral image, birds are positioned in lateral recumbency. The wings are stretched dorsal to the body in a position that the bones of the wings don't project over each other. The wings are fixed to the table or X-ray plate with tape. The upper wing should be in a horizontal plane, as bringing it closer to the table or X-ray plate can cause rotation of the body resulting in incorrect positioning. The legs are pulled ventrocaudally (to minimize superimposition of the leg and the internal organs) and fixed to the table or X-ray plate with tape (**Fig. A3.2**). For examining the legs, one leg should be positioned a bit more cranial to prevent superimposition of the two legs.

Note: Some birds, for example birds with certain orthopedic or anatomical abnormalities of the shoulder girdle, cannot fully stretch the wings above the body. Trying to do this anyway with force can lead to fractures.

VENTRODORSAL PROJECTION

For the ventrodorsal images, birds are positioned in dorsal recumbency with the wings symmetrically stretched to the sides and legs symmetrically stretched caudally. The wings and legs can be fixed to the table or X-ray plate with adhesive tape (**Fig. A3.3**). The keel (Carina) should project exactly over the spine (in birds without skeletal malformation).

45° OBLIQUE PROJECTION

In case of trauma to the shoulder girdle, additional 45° oblique X-rays (**Fig. A3.4**) can be made to prevent superimposition of the coracoid and scapula. X-rays in laterolateral and ventrodorsal projection of the body of a correctly positioned bird and schematic anatomy of the internal organs are shown in **Fig. A 3.5–A. 3.8**.

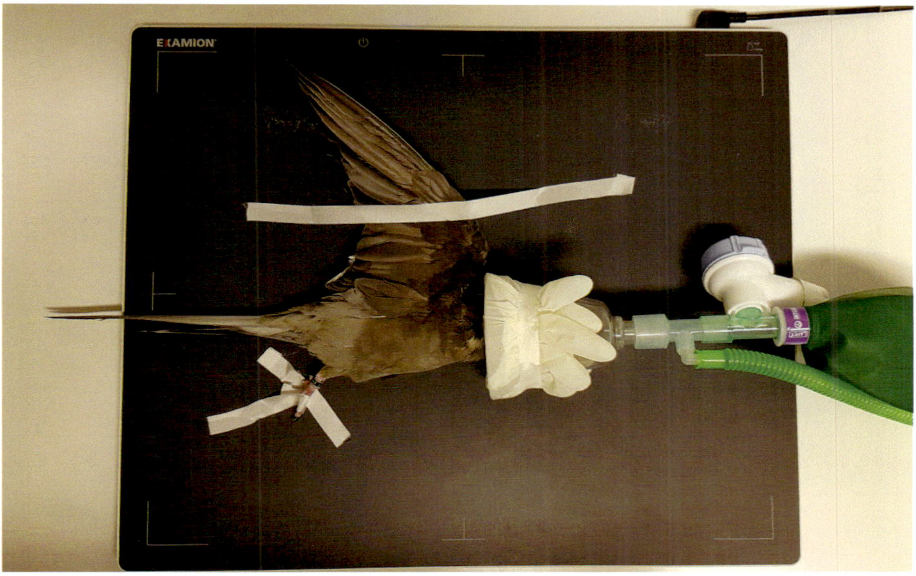

Fig. A3.2 Positioning for laterolateral X-ray. Note the legs being pulled in ventrocaudal direction to minimize superimposition of the upper legs over the internal organs.

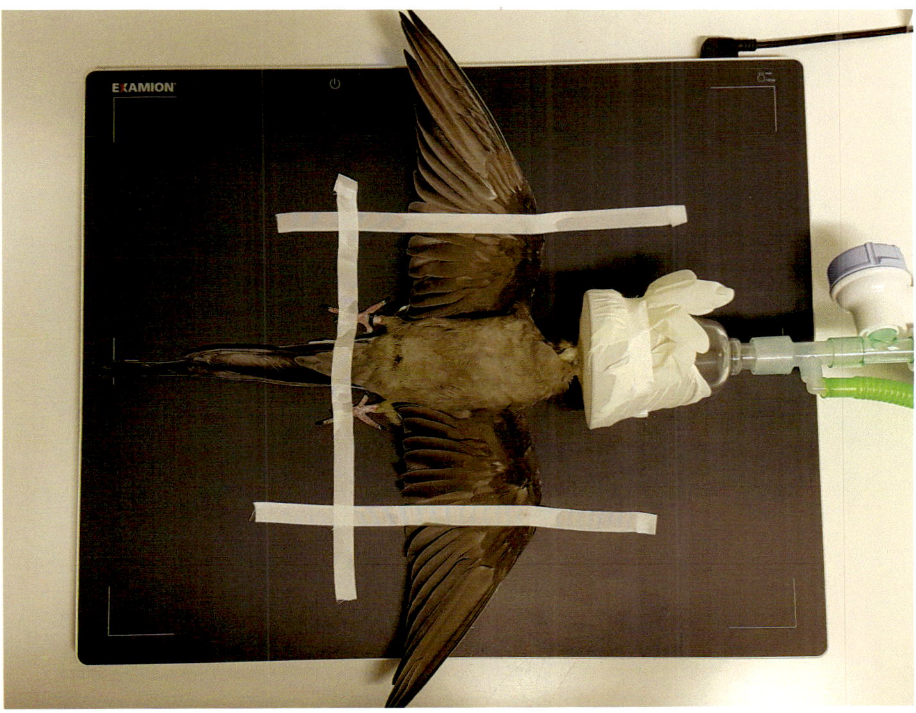

Fig. A3.3 Positioning for ventrodorsal X-ray.

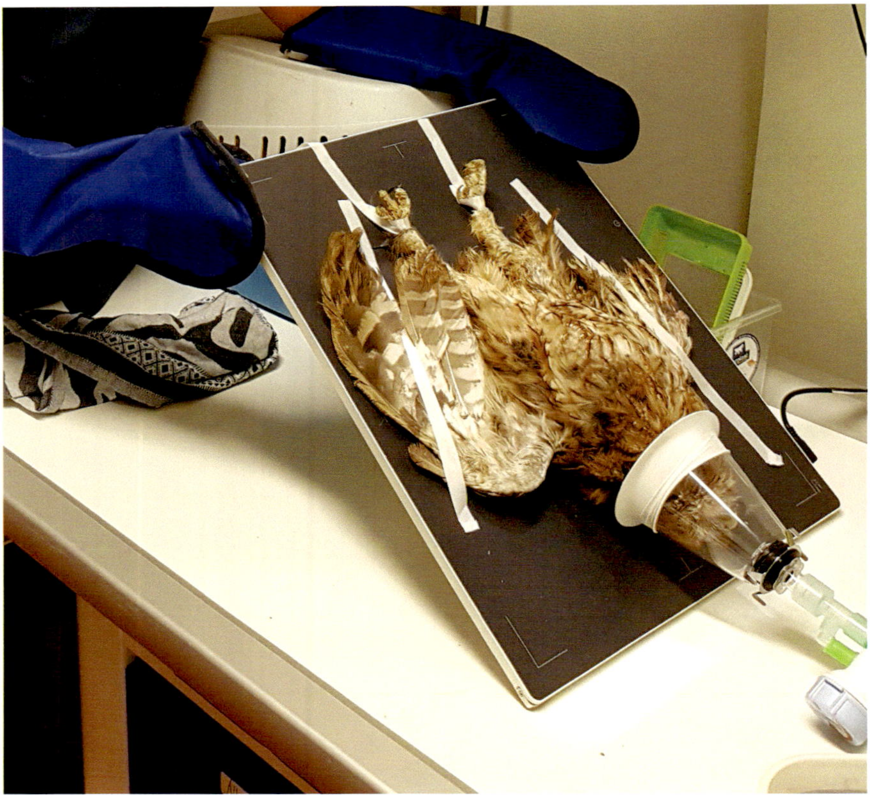

Fig. A3.4 45° oblique X-ray to prevent superimposition of the coracoid and scapula.

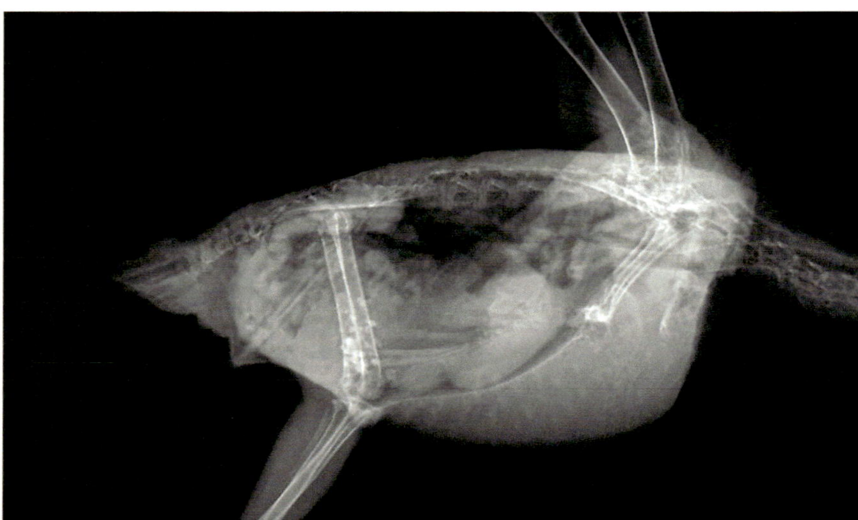

Fig. A3.5 Laterolateral X-ray of a psittacine.

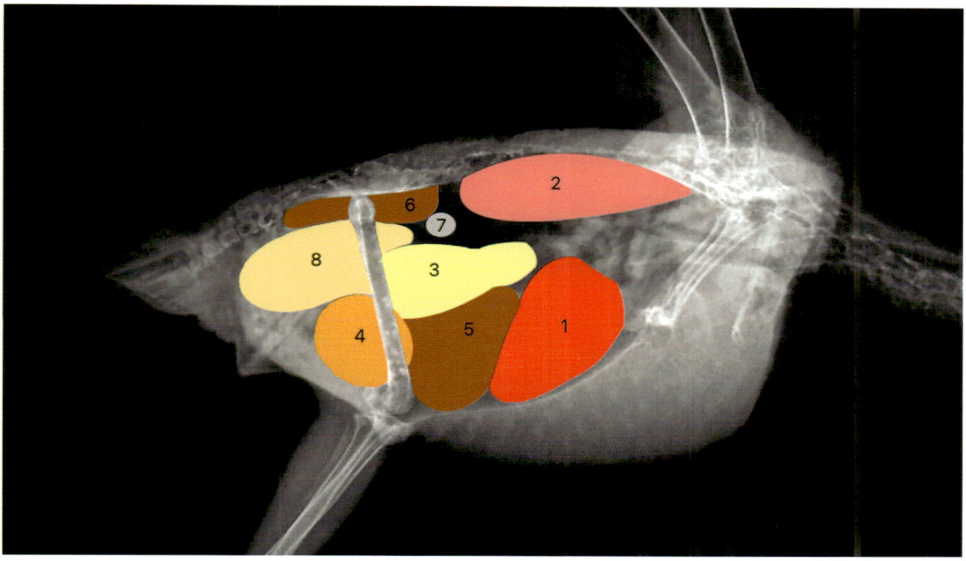

Fig. A3.6 (1) Heart, (2) lungs, (3) proventriculus, (4) ventriculus, (5) liver, (6) kidneys, (7) testes or ovarium and (8) intestines and oviduct/uterus in females.

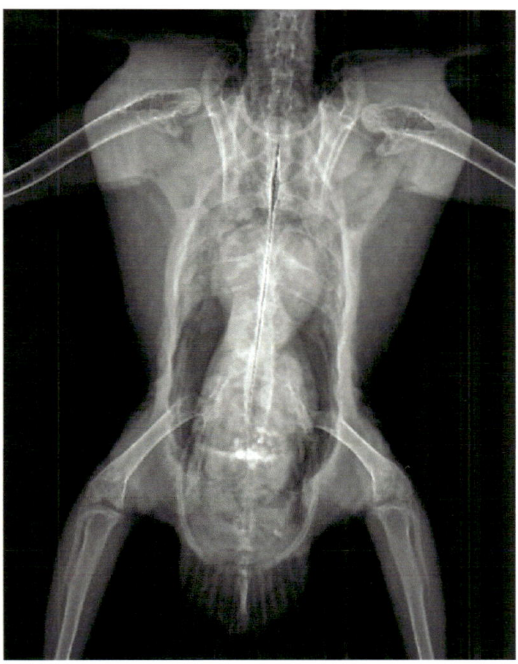

Fig. A3.7 Ventrodorsal X-ray of a psittacine.

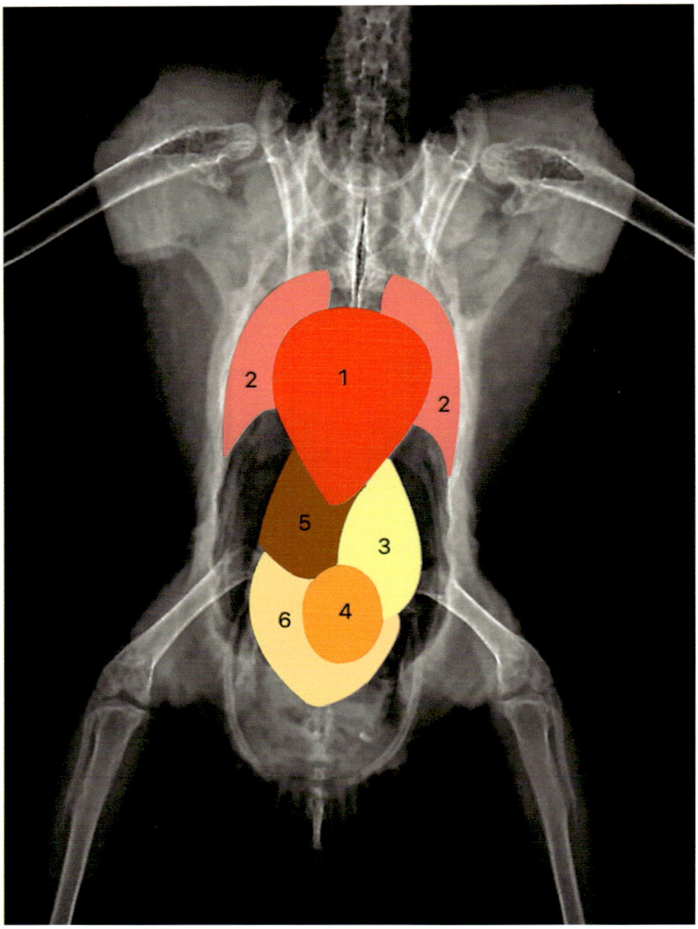

Fig. A3.8 (1) Heart, (2) lungs, (3) proventriculus, (4) ventriculus, (5) liver and (6) intestines, kidneys and oviduct/uterus in females.

CONTAINER

If, due to the patient's physical condition or other factors (like absence of anesthetics), it is not possible to safely take correctly positioned X-rays during the emergency consultation, in certain cases it may be decided to take an incorrectly positioned but usable X-ray. For this, birds can just sit on the table or X-ray plate in a radiolucent plastic container (**Fig. A3.9**) while an X-ray is made with a vertical radiation beam. An alternative is to take a laterolateral X-ray with a horizontal radiation beam while the bird is perched.

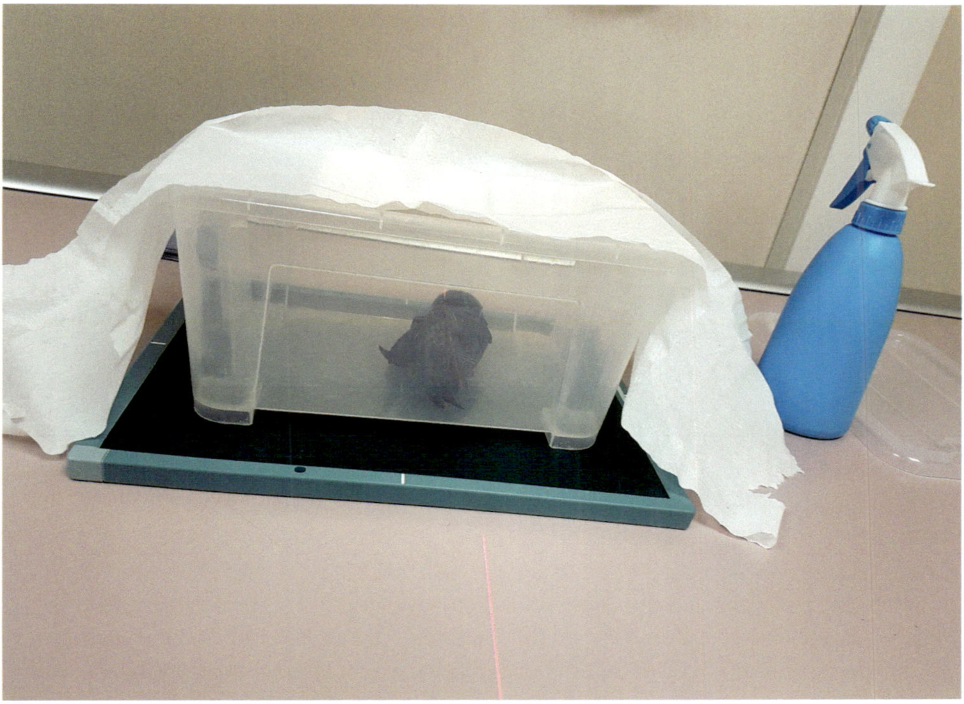

Fig. A3.9 X-ray of a bird in a radiolucent container.

X-rays made in these ways are usually not suitable for assessing internal organs (with the exception of the front part of the gastrointestinal tract), as positioning is not optimal and organs (and limbs on laterolateral X-rays) are superimposed. However, these X-rays are without the risks of anesthesia and they can be useful in emergency situations for determining the presence of for example metal particles (**Fig. A3.10**), radiodense foreign bodies, calcified eggs (**Fig. A3.11**), certain fractures and for examining skeletal calcification.

Fig. A3.10 X-ray of a galah sitting in a radiolucent container. Despite suboptimal positioning, metal particles (arrow) are clearly visible in the ventriculus.

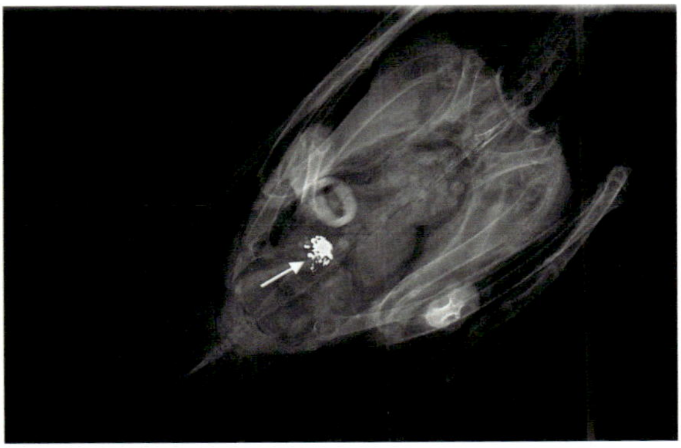

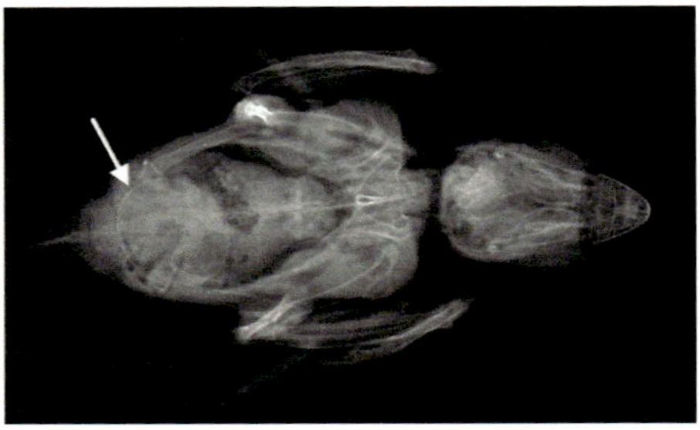

Fig. A3.11 X-ray of a pyrrhura sitting in a radiolucent container. Despite suboptimal positioning, a thin shelled egg (arrow) is visible in the caudal coelom. There is no increased radiodensity of the long bones (hyperostosis), indicating that calcium deficiency could be the cause of the egg binding in this case.

CONTRAST STUDIES

Contrast studies are very useful in avian practice. Not only for diagnosing abnormalities of the gastrointestinal tract, but also for getting a better view of the size and position of other organs and masses in the coelom. Especially in cases of ascites, obesity and big masses, differentiating between the different structures in the coelom can be very difficult on normal X-rays. Displacement of the gastrointestinal tract (clearly visible on contrast studies) can be an indicative of what organ is enlarged or where masses are located in the coelom. For example, hepatomegaly causes a caudodorsal displacement of the gastrointestinal tract and enlargement of testis, ovary, oviduct or kidney cause a ventral displacement of the gastrointestinal tract. See examples later in this appendix.

For contrast studies, barium sulfate (20ml/kg, diluted 1:1 or 1:2 with water) is introduced into the crop (or proventriculus in case of owls) through a crop tube. X-rays are made every 30 minutes with the bird sitting awake in a radiolucent container to check the location of the contrast medium. Transit time through the gastrointestinal tract varies between species and individuals, especially in sick birds. Ideally, the correctly positioned X-rays are made when the contrast medium fills up the entire gastrointestinal tract except the crop and cervical oesophagus.

Note: Contrast medium present in the crop and cervical oesophagus can lead to regurgitation and aspiration in sedated birds not being held in an upright position with the head up!

EXAMPLES OF X-RAY STUDIES

Figs. A3.12-A3.27 show examples of pathological changes on X-rays.

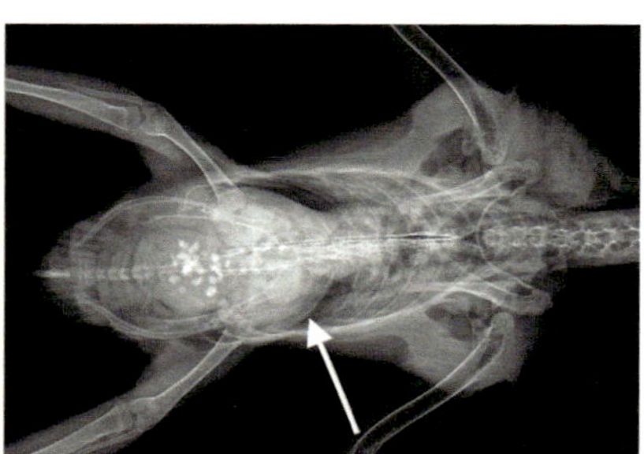

Fig. A3.12 Enlarged proventriculus (arrow).

Fig. A3.13 Enlarged proventriculus (between arrows).

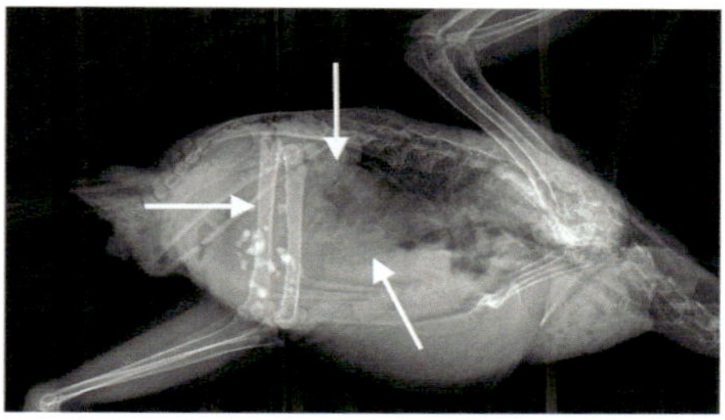

Fig. A3.14 Radiodense mass (between arrows) in the lung of a parakeet.

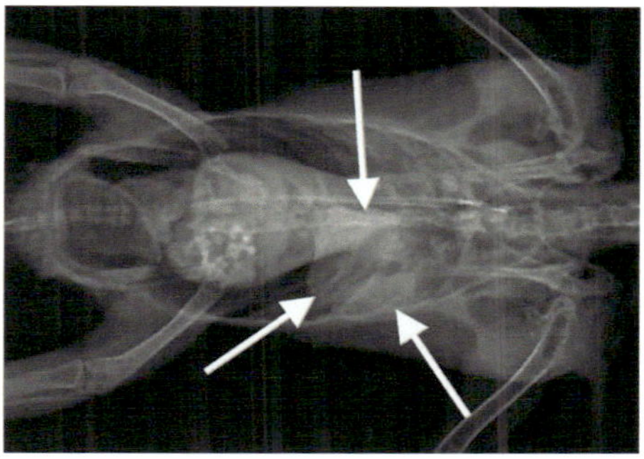

Fig. A3.15 Swollen, radiodense kidneys (between arrows) in a lovebird.

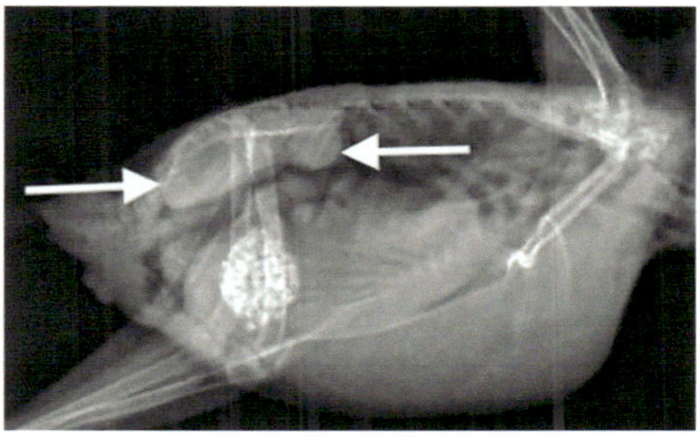

Examples of contrast studies

Fig. A3.16 Enlarged proventriculus filled with contrast medium.

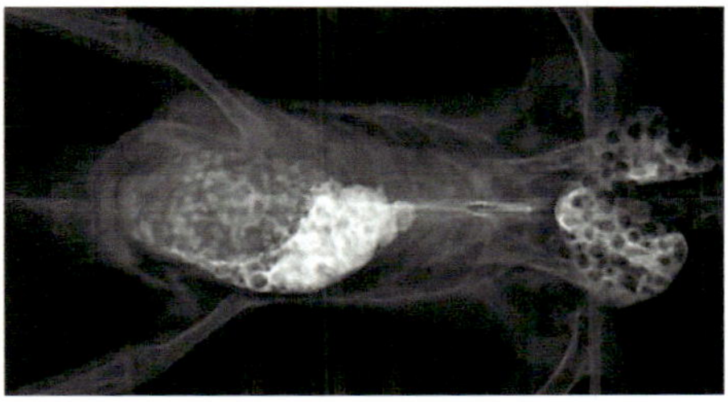

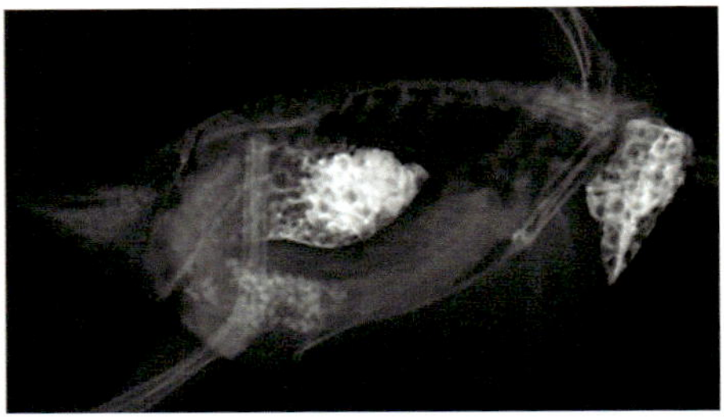

Fig. A3.17 Enlarged proventriculus and dilated intestines filled with contrast medium.

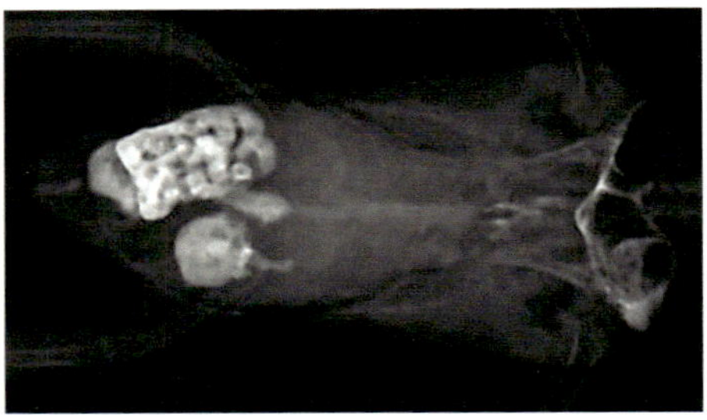

Fig. A3.18 The gastrointestinal tract is displaced in a caudal direction by an enlarged liver.

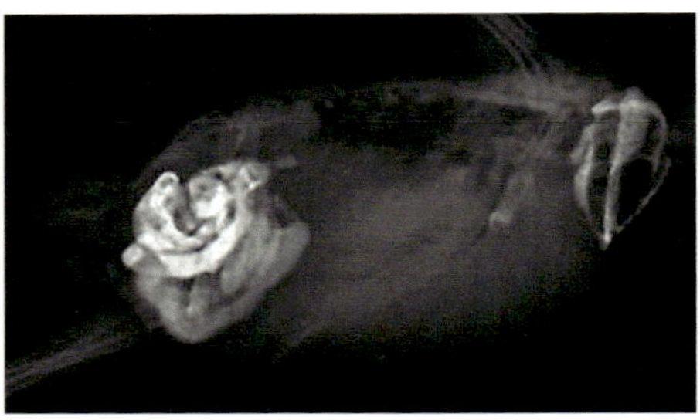

Fig. A3.19 The gastrointestinal tract is displaced in a caudodorsal direction by an enlarged liver.

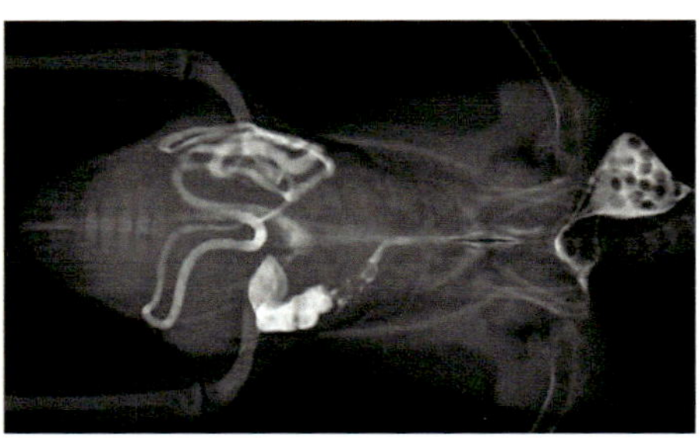

Fig. A3.20 The proventriculus and ventriculus are displaced in a craniolateral direction by a mass (in this case an enlarged oviduct).

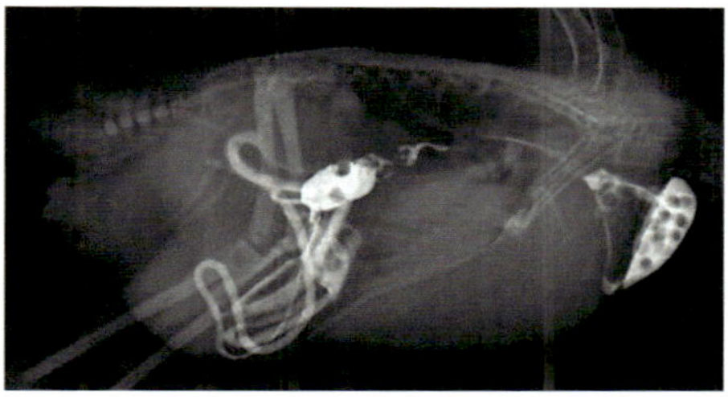

Fig. A3.21 The intestines are displaced in a cranioventral direction by a mass (in this case an enlarged oviduct).

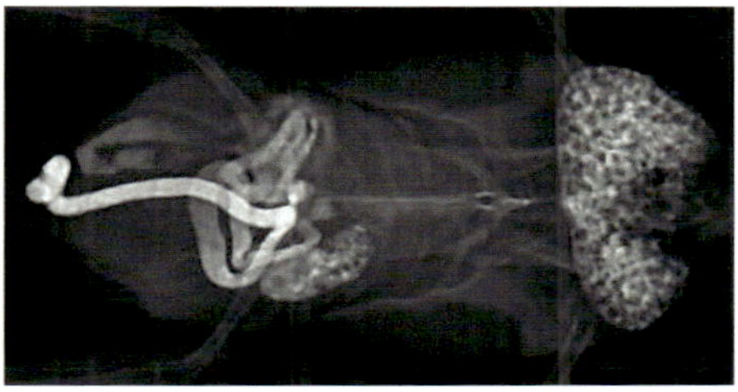

Fig. A3.22 The gastrointestinal tract is displaced in a cranial direction by a mass (in this case an ectopic egg without shell calcification free in the coelom).

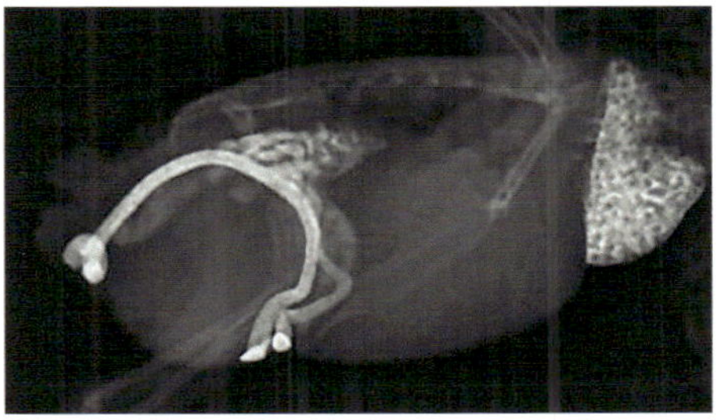

Fig. A3.23 The intestines are displaced in a craniodorsal direction by a mass (in this case an ectopic egg without shell calcification free in the coelom).

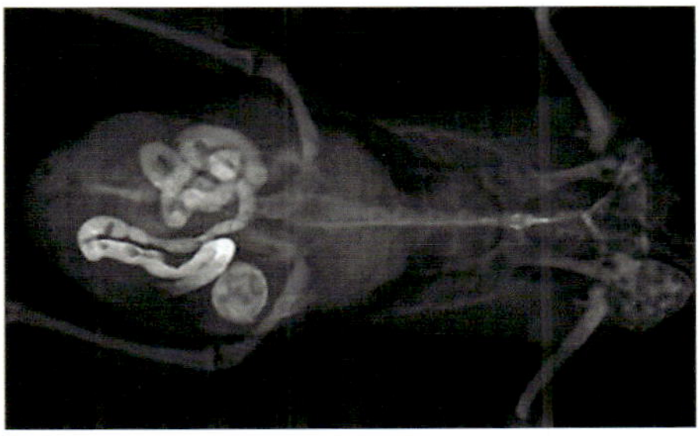

Fig. A3.24 Soft tissue mass outside the gastrointestinal tract (in this case, an enlarged testicle). Polyostotic hyperostosis is visible in the long bones (in male birds, this is indicative of testicular neoplasia).

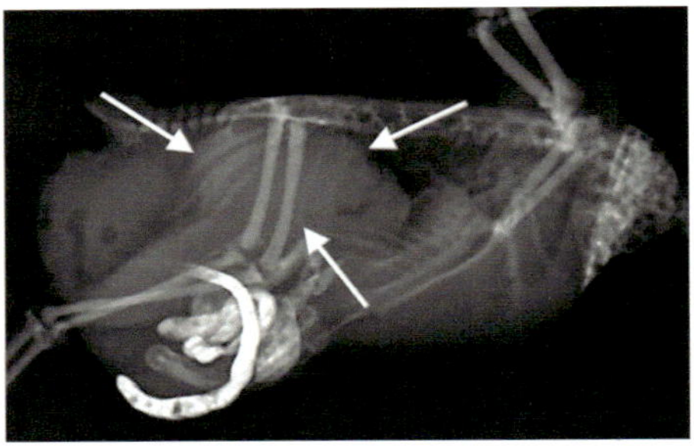

Fig. A3.25 The gastrointestinal tract is displaced in a ventral direction by a mass (between arrows; in this case, an enlarged testicle). Polyostotic hyperostosis is visible in the long bones (in male birds, this is indicative of testicular neoplasia).

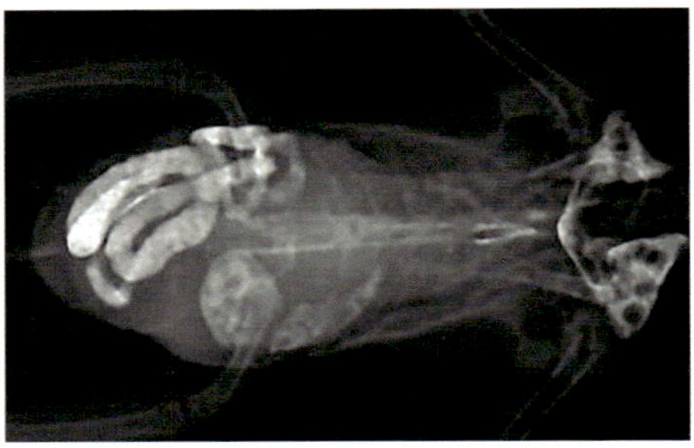

Fig. A3.26 Soft tissue mass outside the gastrointestinal tract (in this case, an enlarged kidney). No polyostotic hyperostosis.

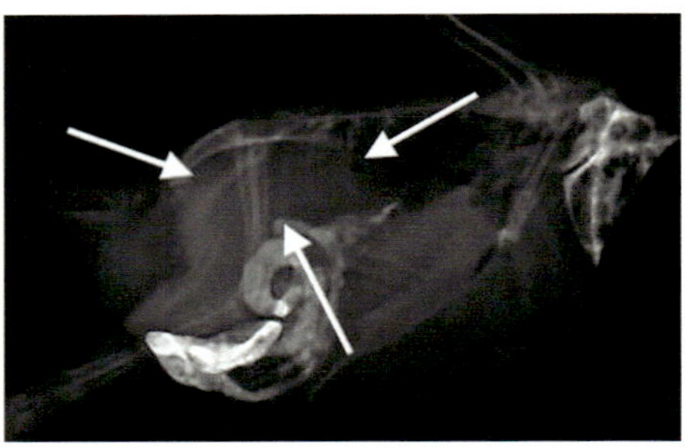

Fig. A3.27 The gastrointestinal tract is displaced in a ventral direction by a mass (between arrows; in this case an enlarged kidney). No polyostotic hyperostosis.

A4 – Microscopic examination of feces

WET MOUNT

A sample of fresh feces is diluted with a small amount of warm saline. A few drops of the mixture are placed on a microscope slide and covered with a coverslip. The coverslip area is scanned for worm eggs (**Fig. A4.1**), *Macrorhabdus ornithogaster* (avian gastric yeast), larvae, flagellated protozoans, protozoan cysts and oocysts (**Fig. A4.2**) with a 10× objective and the diaphragm partially closed. Higher magnification can be used to have a better view of suspicious objects/organisms.

Note: Parasitic infections are quite common in free ranging birds (in many cases not clinically relevant) and outdoor aviary birds (frequently clinically relevant), but uncommon in birds living indoors (usually clinically relevant).

Fig. A4.1 Worm egg (*Capillaria*).

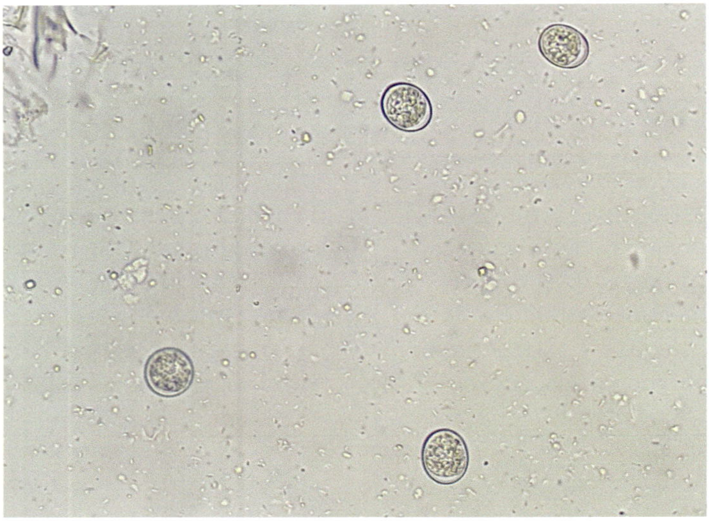

Fig. A4.2 Coccidial oocysts.

STAINED SMEAR

A thin smear of feces is dried and stained using quick stains (e.g., Hemacolor) or Gram stain. The smear is examined with the 40× and 100× objective and the diaphragm opened. Stained smears can be examined for the presence of yeasts (e.g., *Macrorhabdus ornithogaster*, **Fig. A4.3**; *Candida*, **Fig. A4.4**; bacteria, **Fig. A4.5**; inflammatory cells and red blood cells, **Fig. A4.6**).

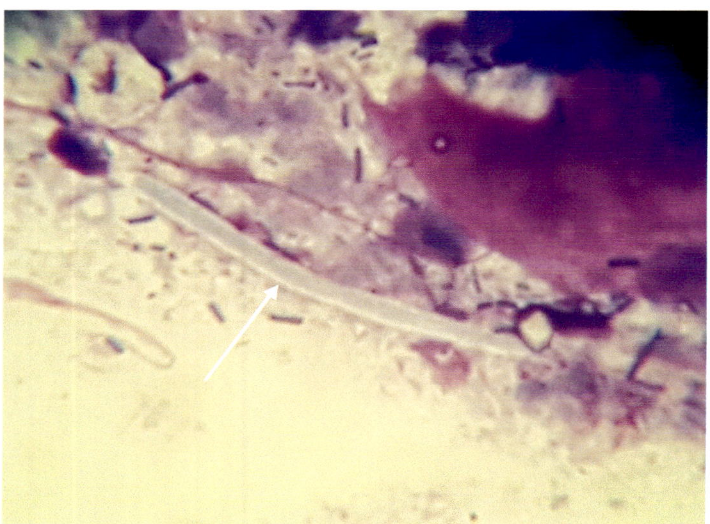

Fig. A4.3 *Macrorhabdus ornithogaster* (arrow).

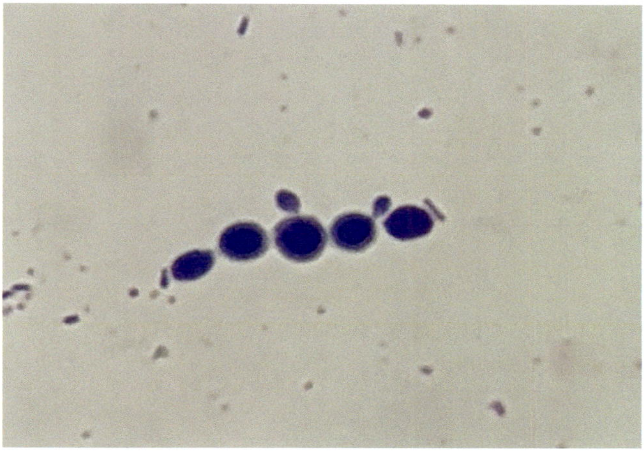

Fig. A4.4 Budding yeast.

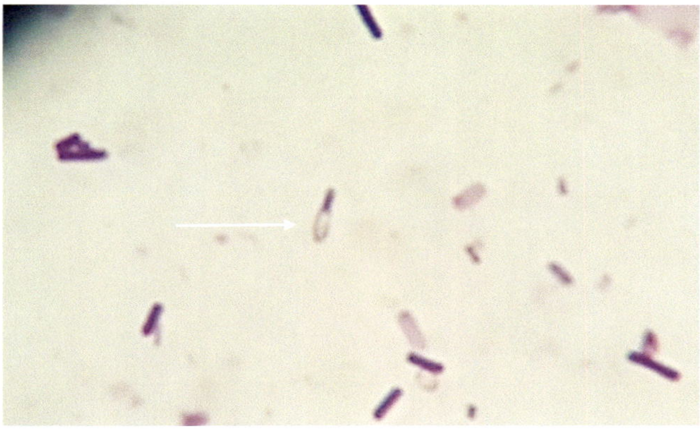

Fig. A4.5 Clostridium with endospore (arrow).

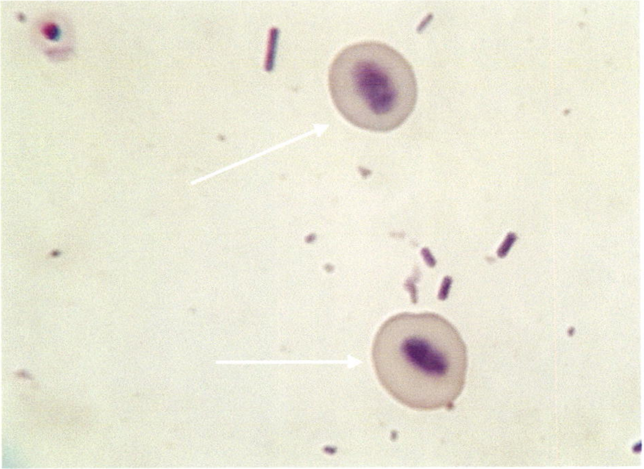

Fig. A4.6 Erythrocytes (arrows) and rod-shaped bacteria.

A5 – Technique: Placement of crop tube and crop lavage

In a neutral position, the neck and esophagus of birds are in a S-shaped position. For pain-free and safe placement of crop tubes, straightening the neck and oesophagus is important. When the handler is also the person placing the crop tube, birds are held with the non-dominant hand. Larger birds are sometimes better held by a second person. The bird is held in an upright position with the head up, facing the person placing the crop tube. The head of the bird is restrained by holding it with light pressure applied with the thumb and middle finger or ring finger to the lower mandible. As described in Appendix A1, the wings and body are wrapped in a towel to prevent moving without pressure on the crop or body. The tube is moistened with some water. A very small amount of lubricant can also be used, but excessive amounts of lubricant can cause aspiration as the tube passes the larynx first. The tube is inserted on the left side of the corner of the beak (**Fig. A5.1**) and then moved over the tongue while staying in contact with the palate to the right side in the back of the oropharynx. Psittacines have a very strong tongue and tend to use it to push the tube out. Making a rolling movement ('top spin') with the tube can facilitate passing the tongue. When the tip of the tube is caudal on the right side of the oropharynx, it is gently directed into the cervical oesophagus, which is located on the right side of the neck (**Fig. A5.2**). If placed correctly in a bird with a straight neck, the tube should slide all the way down into the crop without any resistance. After correct placement, the tip of the tube is easily palpated through the thin skin and wall of the crop and can be felt distinctly separately from the trachea (**Fig. A5.3**). Checking correct placement is essential, as introducing the tube and or fluids into the trachea can lead to death. In most cases, introducing the tube into the trachea leads to sudden resistance or panic of the bird and, of course, dyspnea.

Note: In case of sudden resistance, panic or dyspnea or when in doubt of correct placement, the tube should immediately be removed.

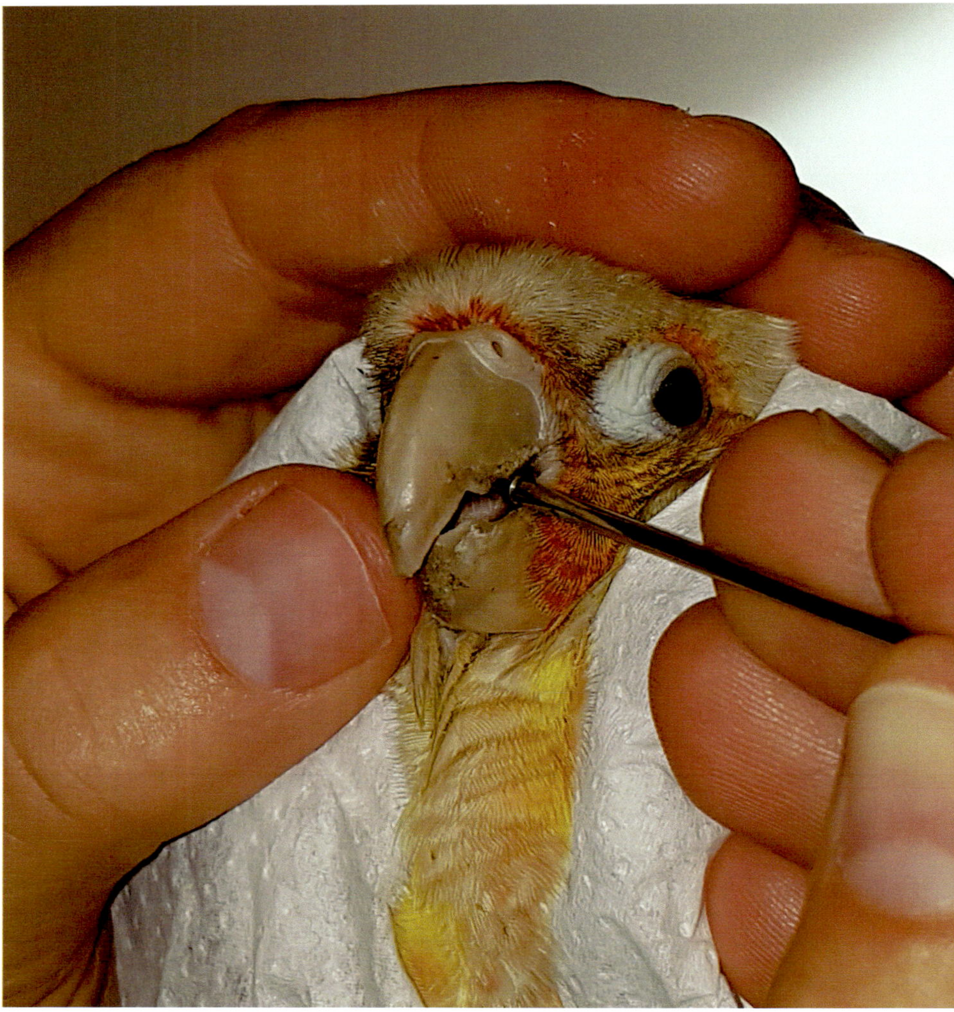

Fig. A5.1 The tube is inserted on the left side of the corner of the beak.

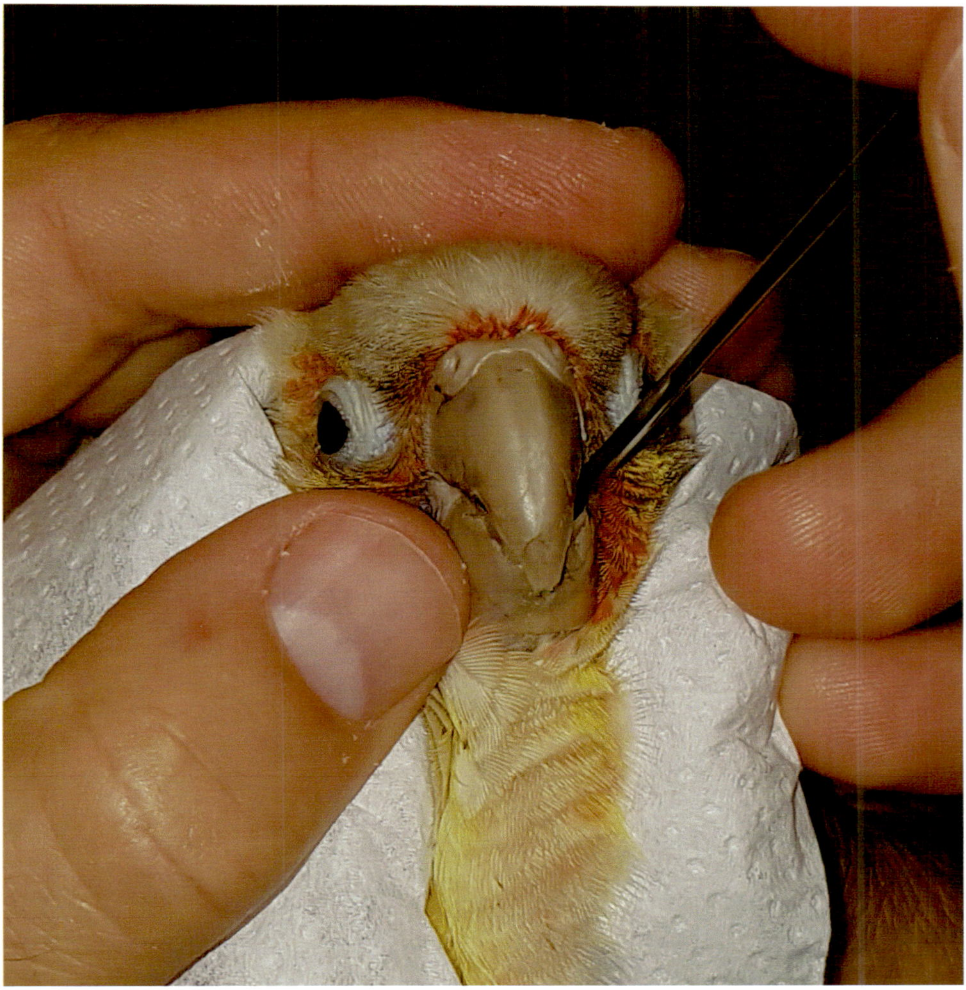

Fig. A5.2 The tip of the tube is moved to the right side of the caudal oropharynx and then directed into the oesophagus.

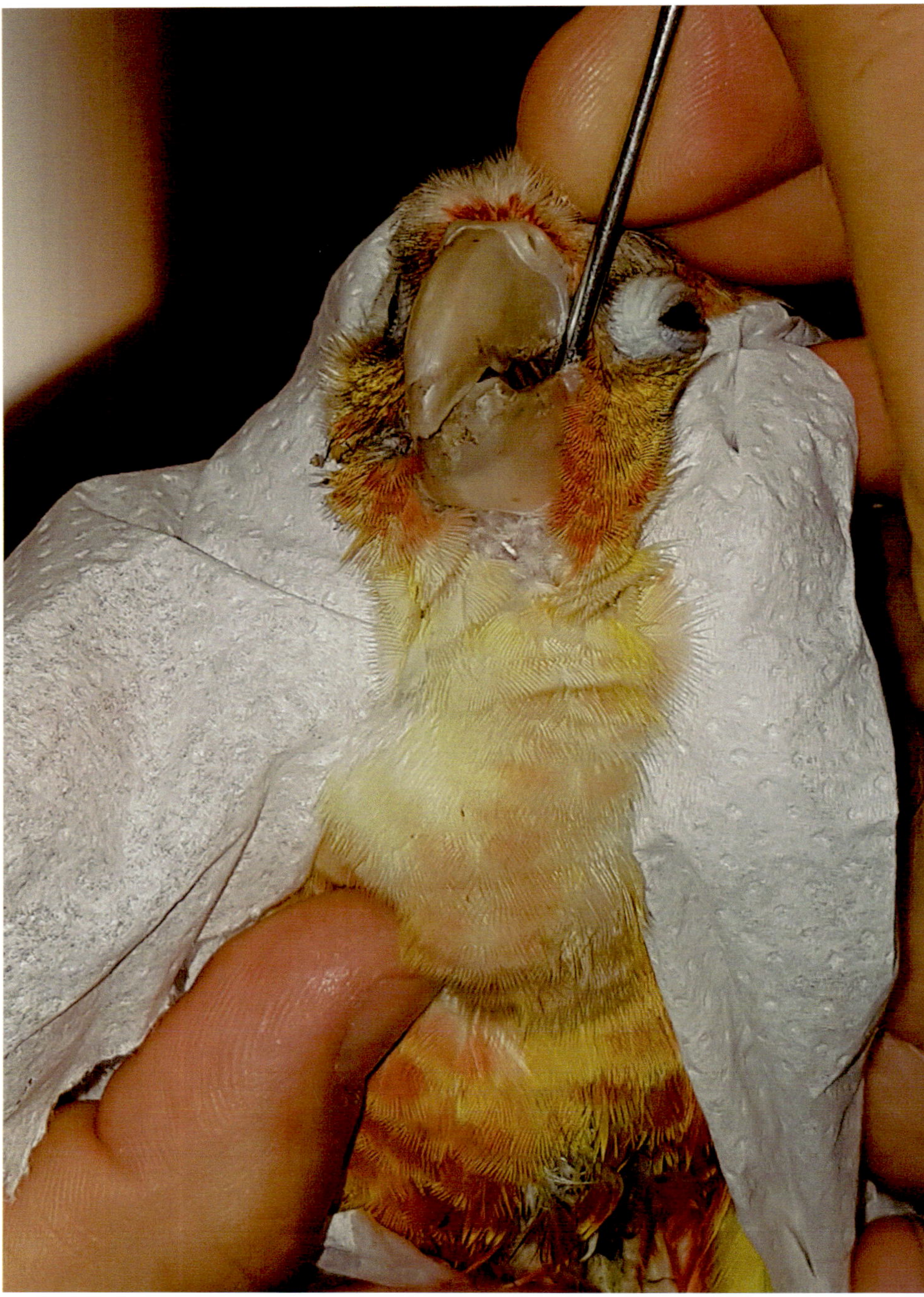

Fig. A5.3 After correct placement, the tip of the tube is easily palpated through the skin and the wall of the crop and can be felt distinctly separately from the trachea. In this photo, palpation is done with the thumb in the middle of the picture.

After correct placement of the crop tube, liquids (food, medication, electrolytes, water, etc.) can be introduced gently. To prevent aspiration, continuously checking for regurgitation by looking in the oral cavity is indicated while administering liquids. If fluids appear in the oropharynx, the crop tube should immediately be removed and the bird returned to its cage so it can swallow, cough and shake to prevent (further) aspiration.

MATERIAL

Different types of tubes can be used. For most small birds and psittacines, straight or curved metal probes (**Fig. A5.4**) are most practical. For large non-psittacine birds (e.g., waterfowl and chickens) long flexible tubes are suitable. Flexible tubes can be used for psittacines as well, but there is a risk of psittacines biting the tube, resulting in obstruction or damaging and swallowing a part of the tube.

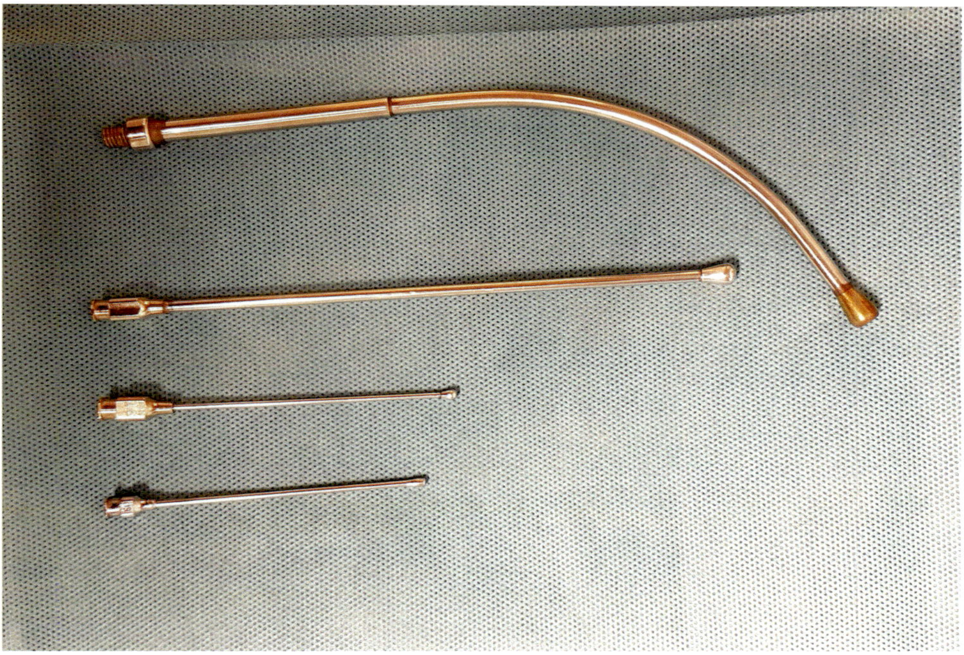

Fig. A5.4 Metal feeding tubes.

Tubes with large diameters cause discomfort and even a risk of damage to the oesophagus. Tubes with very small diameters increase the risk of introduction into the trachea (life-threatening) and perforation of the oesophagus or crop. Ideally, tubes are used that don't fit into the trachea, but are small enough to easily pass the cervical oesophagus.

Examples of suitable tube diameters:

- *African grey parrot/yellow-crowned amazon*: 6 mm
- *Cockatiel/pyrrhura*: 3 mm
- *Budgerigar*: 2–2.5 mm
- *Canary*: 2 mm.

CROP LAVAGE

Lavaging the crop is done by a veterinary professional with or without anesthesia (e.g., with isoflurane/sevoflurane). When done with anesthesia, it's best to intubate the trachea first with an endotracheal tube to prevent aspiration of crop contents before starting the lavage. For crop lavage, a flexible or rigid crop tube is inserted into the crop.

When performed in a bird that is not sedated and intubated, the patient should be held vertically with the head up and only small volumes of fluid should be administered in the crop to prevent aspiration of crop contents after possible regurgitation. Per 100 grams of body weight, 2 ml of fluid is administered into the crop, followed by aspiration of as much fluid as possible into the syringe and discarding this. This step is repeated several times to remove as much material as possible from the crop.

If bigger objects need to be removed from the crop (e.g., pieces of meat in raptors or big metal particles), crop lavage under general anesthesia can be indicated. In sedated birds, larger volumes or even a continuous flush can be used for crop lavage after intubation of the trachea to prevent aspiration of crop contents. The bird can be held horizontally or even with the head slightly down to stimulate emptying of the crop and regurgitation.

Note: To avoid hypothermia, water or NaCl at body temperature should be used for crop lavage.

A6 – Technique: Placement of air sac tube

Air sac tubes can be placed in the caudal thoracic or abdominal air sac to enable breathing in birds with tracheal obstruction. The procedure is stressful and painful and should be performed by a veterinary professional under general anesthesia with adequate analgesia (e.g., isoflurane/sevoflurane or midazolam combined with butorphanol in combination with topical lidocaine).

Depending on the size of the patient, the breathing tube can be a sterile piece of a standard endotracheal tube (**Fig. A6.1**) or, for example, a piece of an IV extension set. If possible, a few extra holes are made on the side at the end of the tube, so that the terminal opening is not the only opening. Needed materials include suture material, needle holders, mosquito forceps, scalpel blade and scissors.

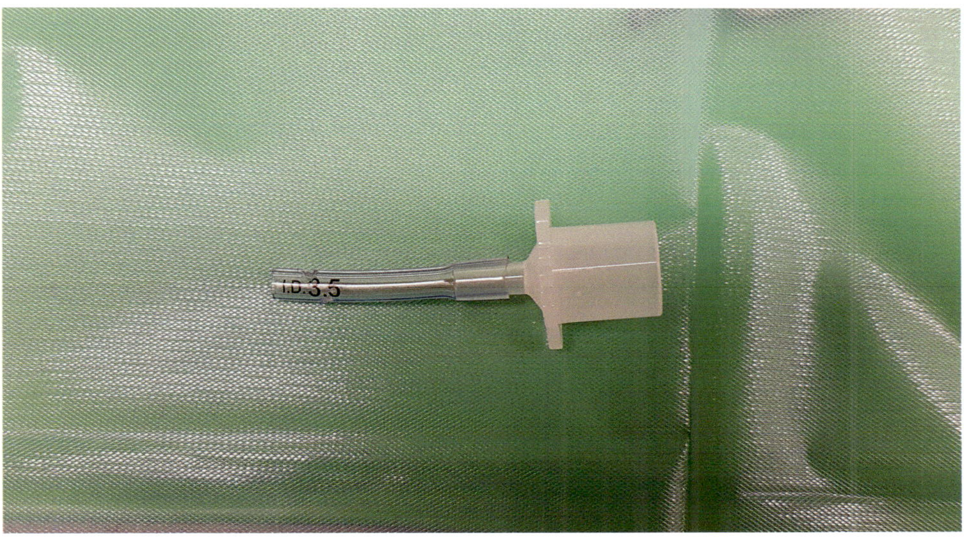

Fig. A6.1 Endotracheal tube cut to length for use as air sac tube.

The bird is placed in a lateral position with the overlying leg pulled cranially (**Fig. A6.2**).

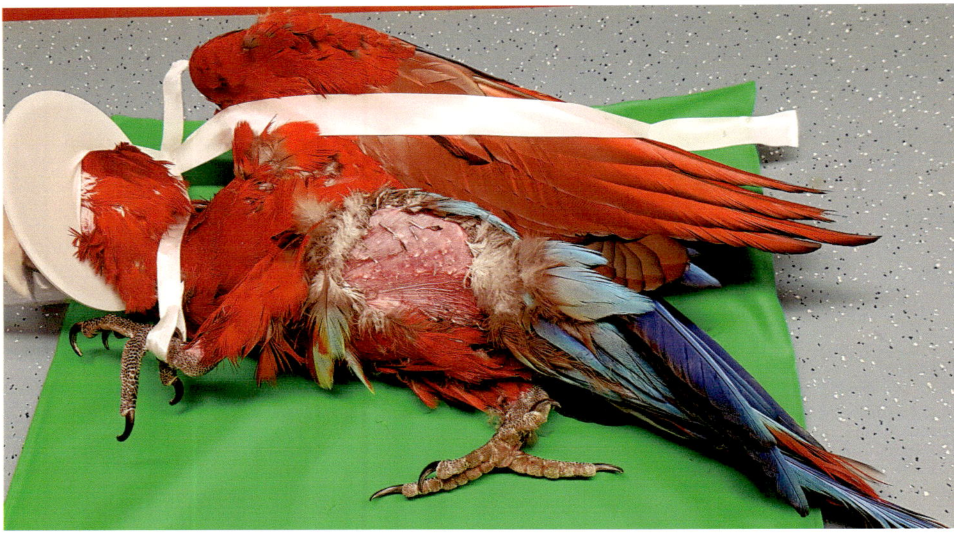

Fig. A6.2 Macaw positioned for placement of air sac tube. Note the upper leg being pulled cranially.

The entry site is in the triangle between the pubic bone, last rib and flexor cruris medialis muscle (**Fig. A6.3**).

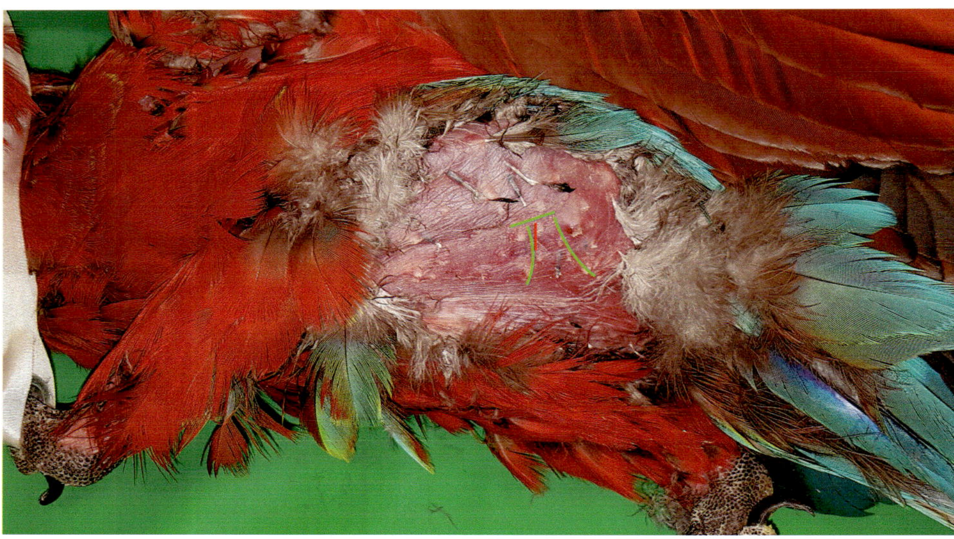

Fig. A6.3 Entry site (red line) just caudal to the last rib, ventral to the M flexor cruris medialis and cranial to the pubic bone (green lines).

The cover feathers overlying the entry site are plucked and the skin is prepared aseptically. A small vertical skin incision is made behind the last rib (**Fig. A6.4**).

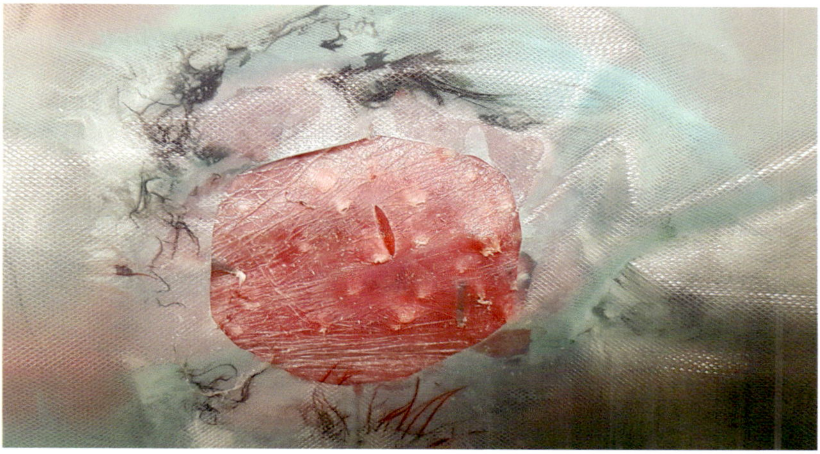

Fig. A6.4 Incision just caudal to the last rib.

The mosquito forceps are used for making a tunnel by blunt dissection dorsally under the flexor cruris medialis muscle first and then medially into to the coelomic cavity (**Fig. A6.5**). The index finger is placed on top of the jaws of the mosquito forceps to prevent damage to internal structures by entering too deep when pushing medially to enter the air sac. Usually a clear 'pop' sensation can be felt or heard when the air sac is entered.

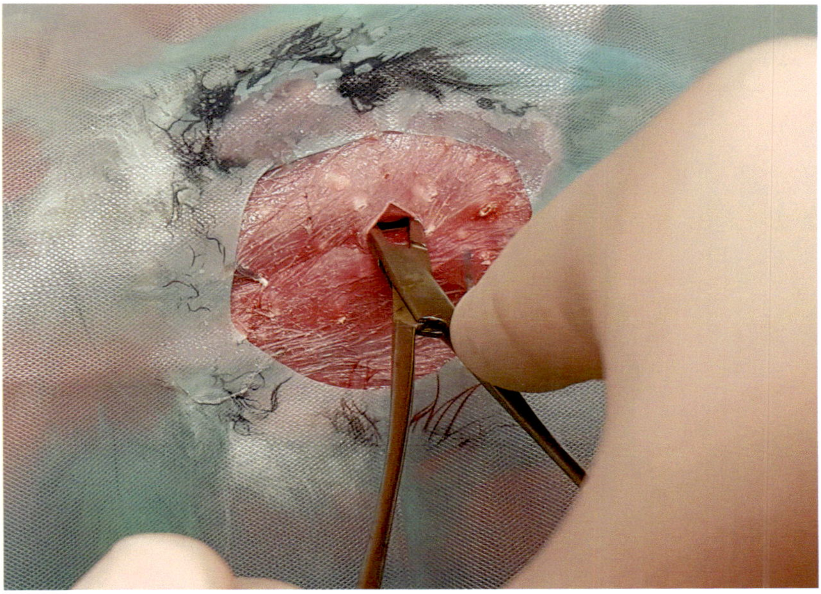

Fig. A6.5 Mosquito forceps are used to make a tunnel into the air sac.

After entering the air sac, the jaws of the mosquito forceps are opened, and the breathing tube can be inserted into the air sac through the jaws of the mosquito forceps (**Fig. A6.6**).

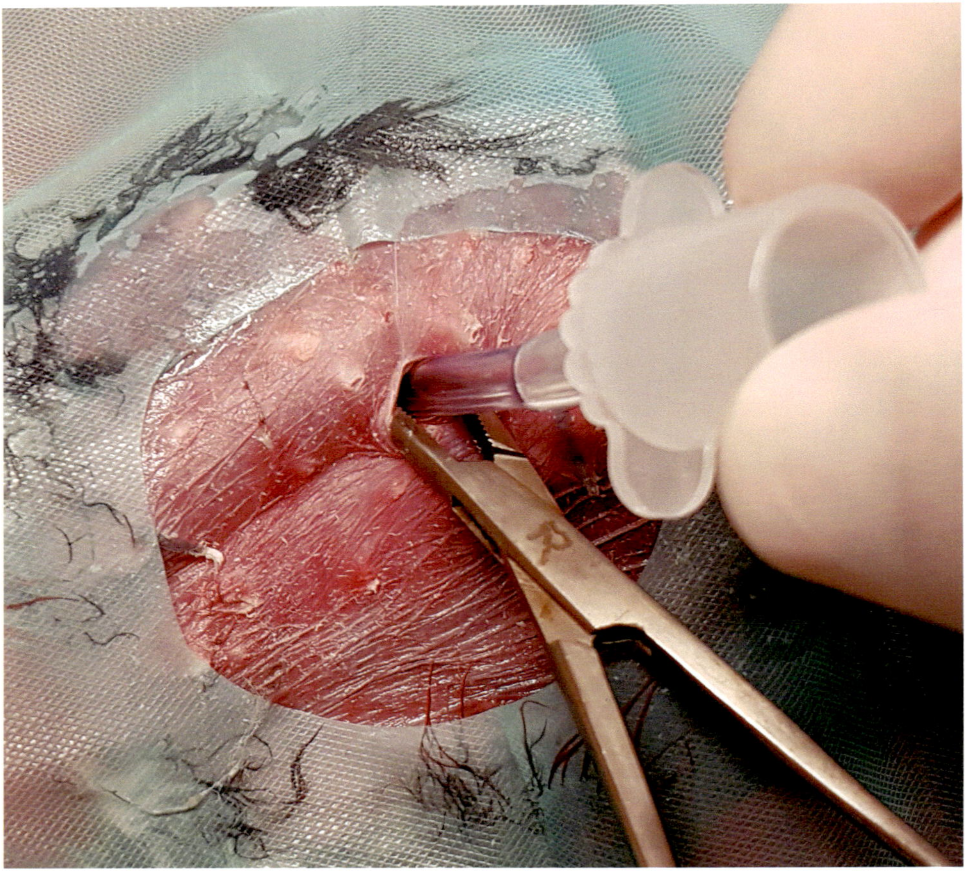

Fig. A6.6 The tube is inserted into the air sac through the opened jaws of the mosquito forceps.

Proper placement will usually result in visible condensation in the tube. Air flow through the tube can be checked by holding a down feather (**Fig. A6.7**) or piece of cotton wool and watching for movement (held tightly to prevent aspiration).

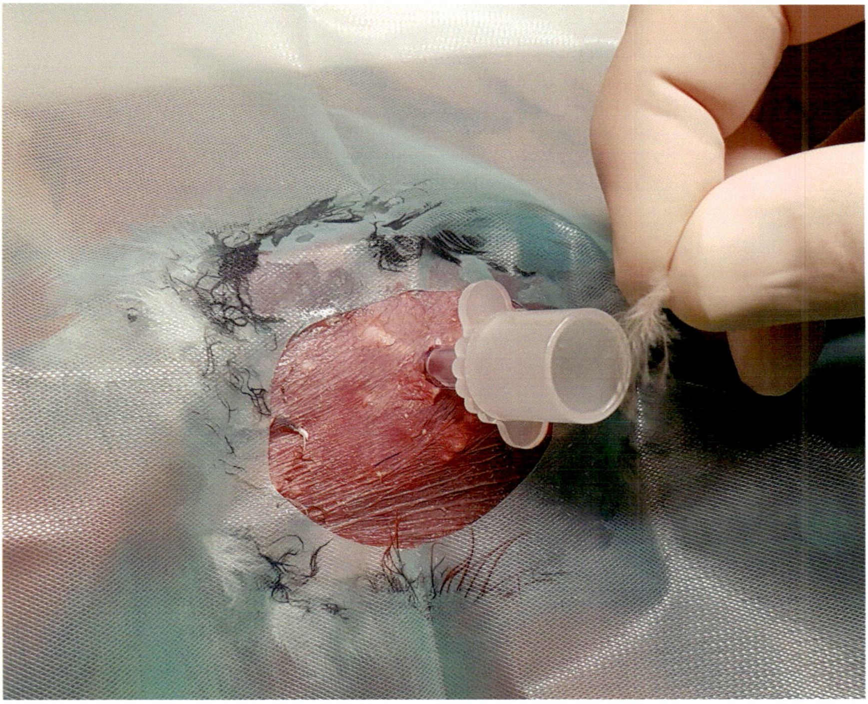

Fig. A6.7 Checking correct positioning and functionality by holding a down feather at the opening of the air sac tube to check the air flow.

The air sac tube is secured in place by attaching it to the skin. This can be done either by applying adhesive tape to the tube and suturing this tape to the skin (**Fig. A6.8**), or by using a finger trap suture.

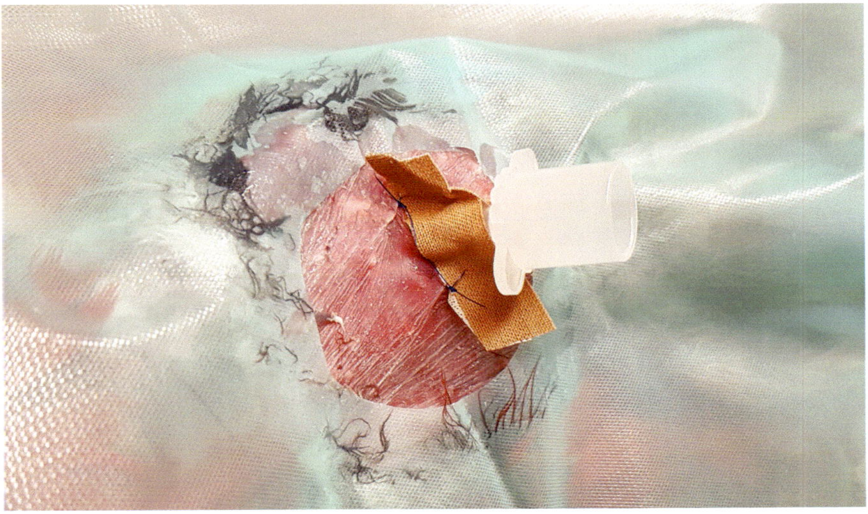

Fig. A6.8 The air sac tube is secured in place.

A7 – Technique: Imploding eggs

Imploding eggs can be indicated in situations of egg binding and should be performed by a veterinary professional. The goal is to remove the contents of the egg and then collapse the shell to alleviate symptoms and to facilitate removal or spontaneous oviposition.

This procedure is done under general anesthesia (e.g., with isoflurane/sevoflurane) with the patient in dorsal recumbency, except when an egg is visible in a prolapsed cloaca. In that situation the procedure can also be done in an awake bird held vertically with the head up.

A right-handed person palpates the egg with the left hand and gently pushes it caudally towards the cloaca. Pressure directed to the spine should be avoided because of the risk of damaging kidneys and nerves. Next the eggshell can be perforated with the needle. In some cases, a small part of the eggshell can be seen in the cloaca and be punctured without damaging soft tissues. When the eggshell is not visible, it is punctured through the mucous membranes of the cloaca. Constant focus on not applying any pressure on the egg directed to the spine while trying to perforate, empty and collapsing is crucial.

For this, the eggshell is perforated with a hypodermic needle. Due to the hardness of the shell and the viscosity of the albumen and yolk, needles with relatively large diameters (for example 21G needles in small psittacines) are best used. The needle is attached to a syringe with a volume a few times bigger than the egg. After perforating the eggshell with a rotating movement (in fully calcified shells, this can be challenging, even in small birds), suction is used to remove the contents of the egg. It may be necessary to move the tip of the needle around a bit inside of the egg and to alternate suction with letting some air back into the egg to get all the albumen and yolk out. After removal of the contents, maximum negative pressure is created by suction. The egg is then collapsed by pressure between thumb and fingers while the vacuum is present, still without applying any downward pressure on the egg. The imploding is often accompanied by a 'pop' sound.

Note: This procedure must not be performed when eggs are not positioned close to the cloaca. Eggs with very thick shells (caused by prolonged egg binding with eggs located in the shell gland, ***Fig A7.1****) cannot successfully be imploded and efforts to do so can result in serious complications.*

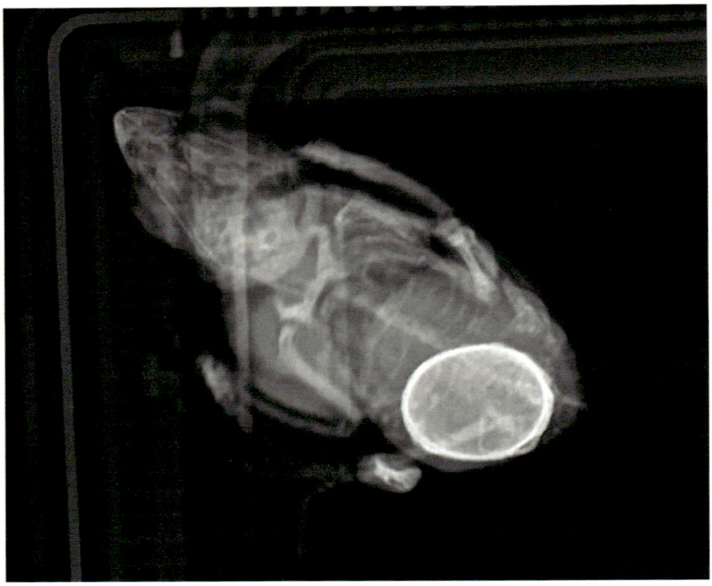

Fig. A7.1 Egg with abnormal thick and irregular shell.

A8 – Technique: Applying (splint) bandages

BODY WRAP

The damaged wing is held against the body in a normal physiological resting position. An elastic bandage is then wrapped around this wing and the body (**Fig. A8.1**), caudal and cranial to the wing on the other side (**Fig. A8.2**).

Note: The bandage must not be tight, otherwise the patient will not be able to breathe properly.

Fig. A8.1 Body wrap viewed from affected side.

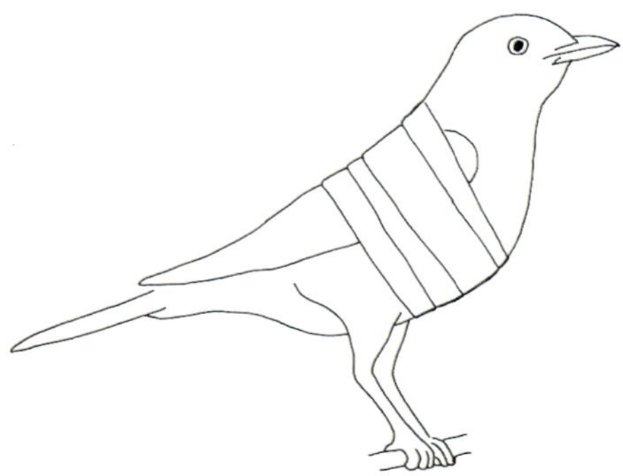

Fig. A8.2 Body wrap viewed from healthy side. The bandage is placed cranial and caudal to the opposite wing to prevent slippage.

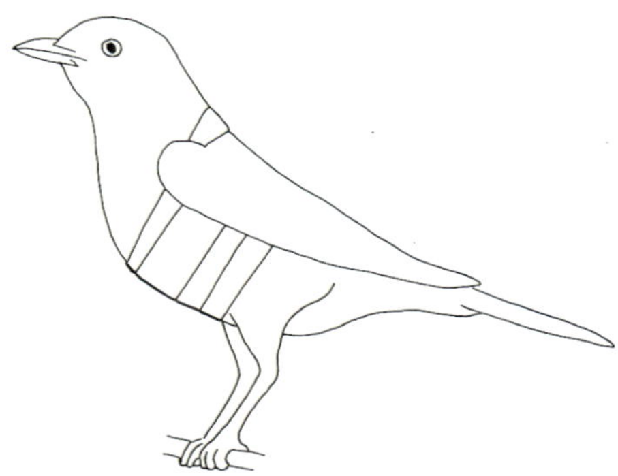

TAPING WING TIPS

This technique can be used for temporarily preventing injured wings to droop. It causes discomfort and is less effective than a body wrap, but can be useful in emergency situations. The most distal two to four primary flight feathers (primaries) from both wings are affixed to each other with adhesive tape (**Fig. A8.3**).

Fig. A8.3 Taping the wing tips together is a simple way to prevent drooping of the wing.

FIGURE-OF-EIGHT BANDAGE

Elastic bandage is wrapped around the bent wrist (left in picture) and then crossed around the distal wing and all the flight feathers,also the flight feathers attached to the upper wing (**Fig. A8.4**). This '8'-shaped bandage provides stability distal to the elbow.

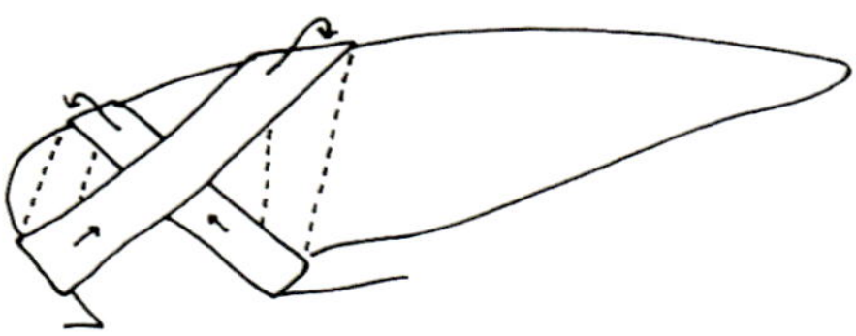

Fig. A8.4 Figure-of-eight bandage.

Note: Prolonged immobilization of the wing can lead to ankylosis and deformities of the propatagium (elastic skin fold extending from the carpal joint to the shoulder on the cranial site of the wing), causing permanent inability to fly. If long-term (>10 days) immobilization is necessary for healing, bandages should be removed every 7–10 days for joint mobilization by passive movement and stretching the propatagium. This is generally done under general anesthesia to prevent complications of the primary problem by active movements of the patient.

TAPE SPLINT LOWER LEG

Pieces of adhesive tape are stuck together from both sides with the leg in the desired position in between (**Figs. A8.5 and A8.6**). Using a needle holder, tweezers or forceps, the layers of tape are firmly pressed together close to the leg (dotted line). Multiple layers of tape provide more stability, but care must be taken that the splint does not become unnecessarily heavy. Excess tape is cut off 0.5–1 cm from the leg (**Fig. A8.7**). A layer of super glue can provide extra strength; this should only be applied if the splint is the final method of fixation, not if it is used as an emergency bandage that needs to be removed again prior to final treatment.

Fig. A8.5 After correct positioning, pieces of tape are adhered to one side of the lower leg.

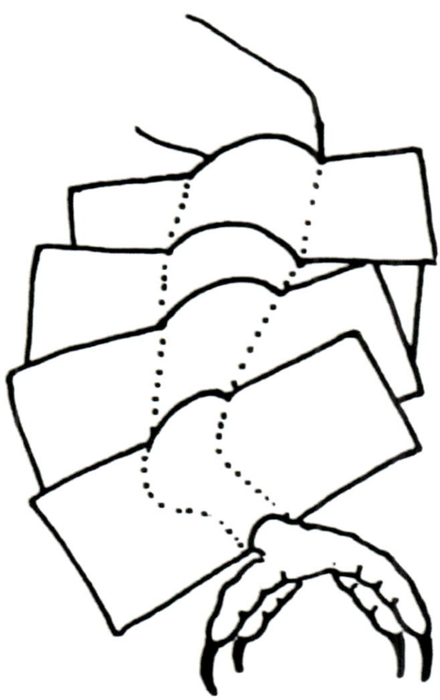

Fig. A8.6 Then, pieces of tape are applied to the other side. The tape fragments from both sides are pressed together close to the leg.

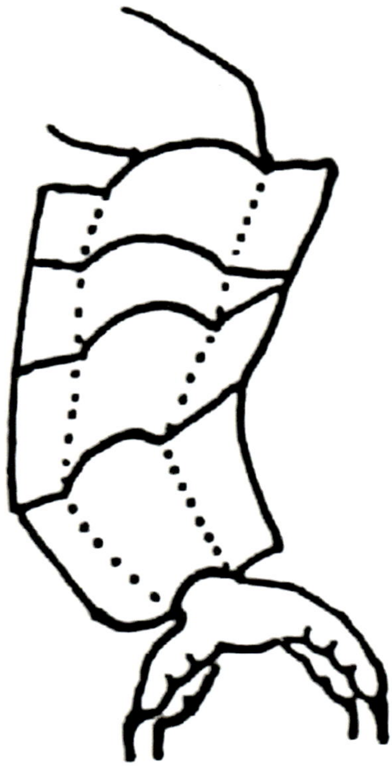

Fig. A8.7 Excess tape is cut off.

A9 – Technique: Ingluviotomy

Ingluviotomy is the surgical opening of the crop. The procedure is done under general anesthesia (e.g., with isoflurane/sevoflurane) with adequate analgesia (e.g., meloxicam and local lidocaine/butorphanol) by a veterinary professional. When possible, an endotracheal tube is placed after induction to prevent aspiration of regurgitated crop contents. The patient is placed in dorsal recumbency and the feathers covering the crop region are plucked (**Fig. A9.1**). After aseptic preparation of the surgical field, the skin is incised (**Fig. A9.2**) followed by an incision through the wall of the crop (**Fig. A9.3**), avoiding the visible blood vessels in the wall of the crop. After removal of abnormal crop contents, the crop wall is closed with monofilament absorbable suture material (5–0, 4–0) in two layers, the first layer with a continuous opposing suture pattern (**Fig. A9.4**), the second layer with a continuous inverting suture pattern (**Fig. A9.5**). The skin is closed separately in a continuous or simple interrupted pattern (**Fig. A9.6**).

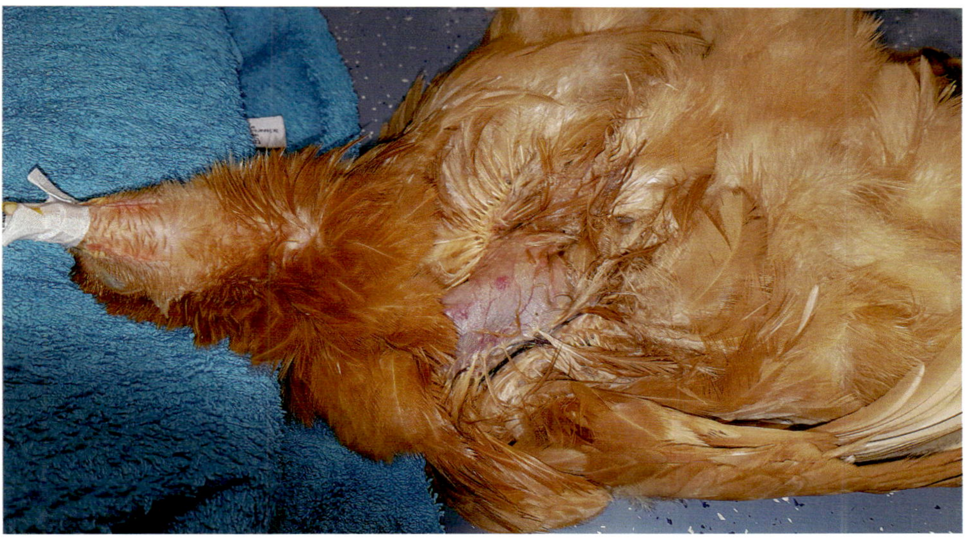

Fig. A9.1 Preparation for ingluviotomy.

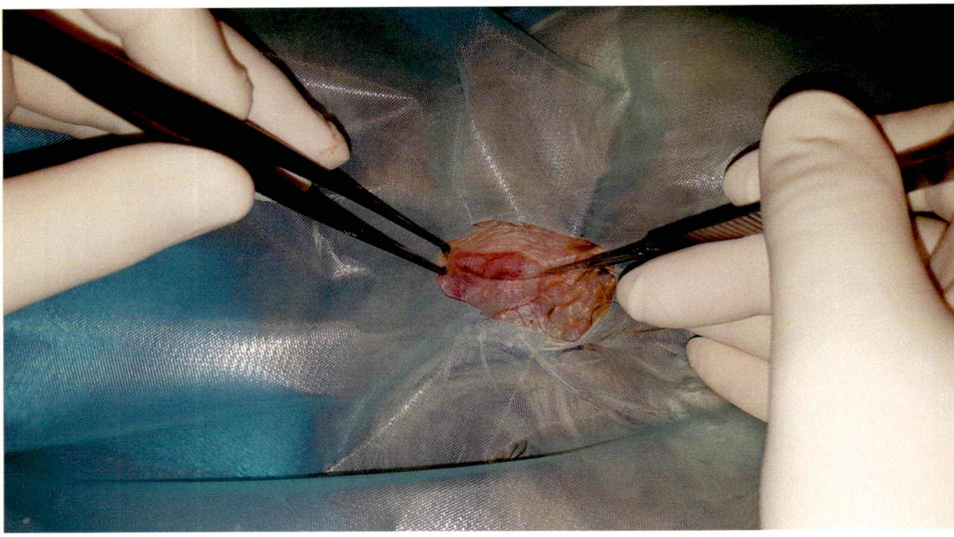

Fig. A9.2 Skin incision.

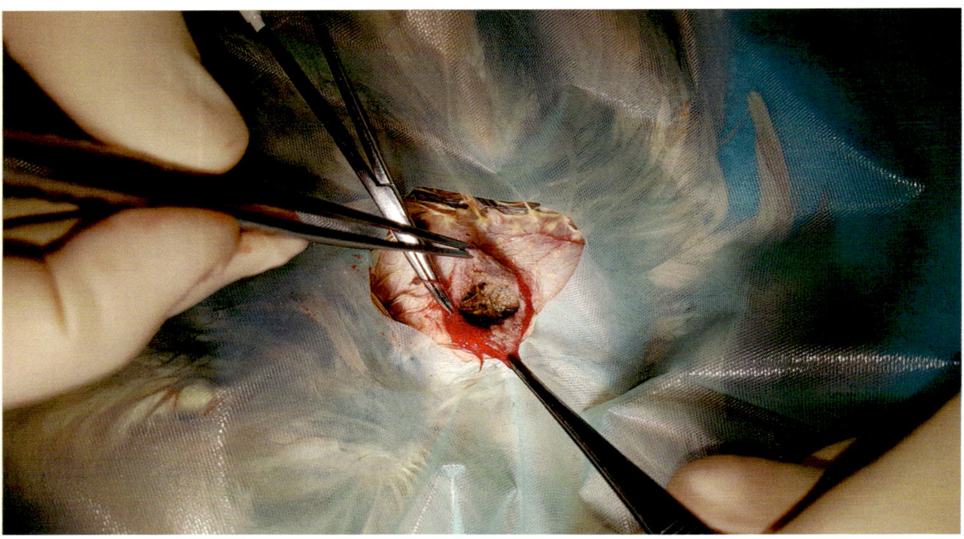

Fig. A9.3 Crop opened.

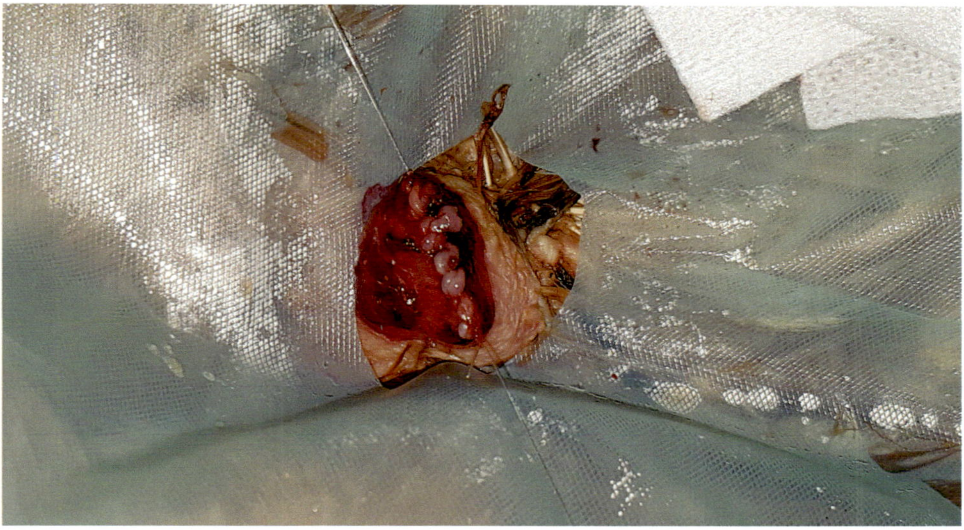

Fig. A9.4 First layer is closed with a continuous opposing suture pattern.

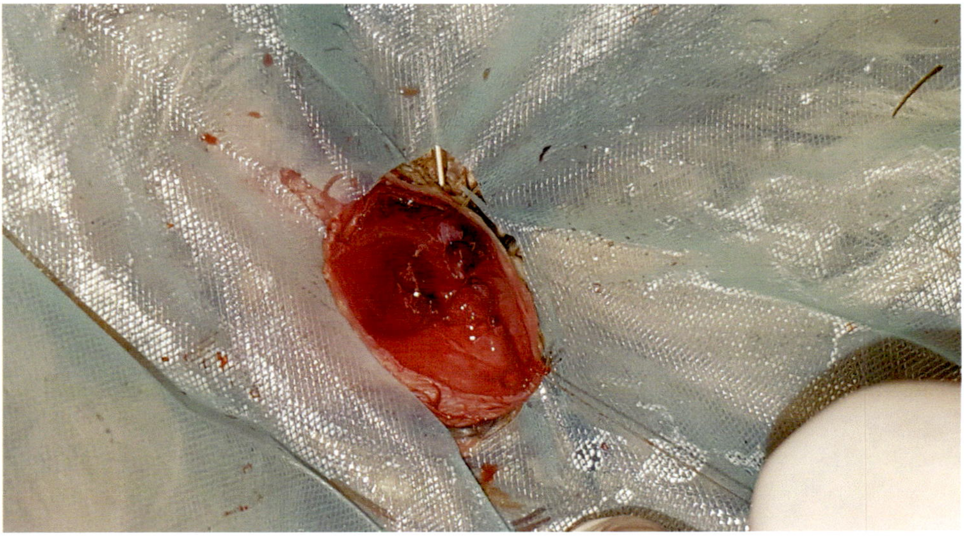

Fig. A9.5 Second layer is closed with continuous inverting pattern.

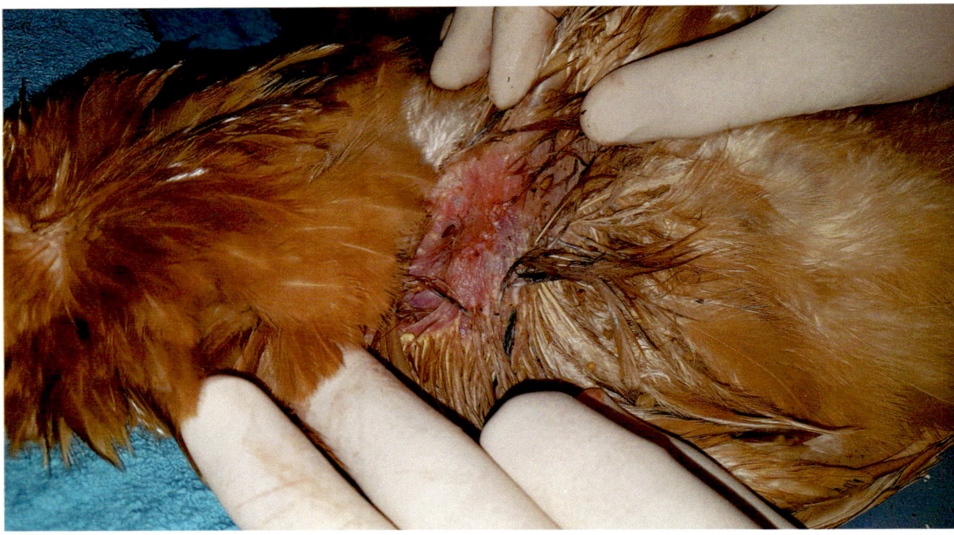

Fig. A9.6 The skin is closed with a continuous or simple interrupted suture pattern.

A10 – Table of (possibly) poisonous plants

Below is a list of plants known to be or suspected of being toxic to birds. The sensitivity of birds to some of the plants on this list is probably overestimated and symptoms after ingestion of parts of some of the listed plant will be mild or even absent. This list will undoubtedly not include every plant toxic to birds and some plants dangerous for birds will not be mentioned.

The plants marked with an asterisk (*) are suspected of serious toxicity. Plants with more than one name can be present multiple times in the list.

*Abrus precatorius**	Jequirity/Precatory/Rosary bean
*Acacia**	Wattle
Acer rubrum	Red Maple
*Aconitum**	Monkshood
Actaea racemosa	Baneberry
Aesculus glabra	Buckeye
Aesculus hippocastanum	Horse Chestnut
Aglaonema Stripes	Chinese Evergreen
*Alocasia**	Elephant ear plant/Taro
Aloe Vera	
Amaranthus	Pigweed
*Amaryllis**	
Amianthium	Fly poison
Ammi majus	Bishop's weed
Andromeda japonica	Japanese andromeda
Anemone	
Anthurium andraeanum	Flamingo flower
Anticlea	Death camas
Antirrhinum majus	Snap Dragon
Araucaria heterophylla	Norfolk Pine
Arctium	Burdock
Argyreia nervosa	Hawaiian baby woodrose
Arisaema triphyllum	Indian Turnip/Jack-In-The-Pulpit
Arum maculatum	Lords-and-Ladies
Asclepias	Milkweed
Asparagus plumosus	Asparagus fern
Asparagus setaceus	Asparagus fern

Astragalus	Locoweed
Atropa belladonna	Belladona
Azalea	Bean plants
Bellis perennis	Daisy
Brachychiton acerifolius	Flame Tree
Brugmansia	Angel's Trumpet
Brunfelsia latifolia	Yesterday, today, tomorrow plant
	Bulb flowers
*Buxus**	Boxwood
*Caesalpinia gilliesii**	Bird of paradise
*Caladium**	Elephant's ear
Callistemon	Bottle Brush
*Caltha palustris**	Marsh-marigold/Kingcup
Campsis radicans	Trumpet Vine
*Cannabis sativa**	Marijuana
Celastrus scandens	Bittersweet
Chrysanthemum	Chrysanth
Cineraria	
Citrus trifoliata	Poncirus
*Claviceps purpurea**	Ergot (fungus growing on rye and related plants)
*Clematis**	
Codiaeum variegatum	Croton
Coffea	Coffee plants
*Colchicum autumnale**	Autumn Crocus/Meadow saffron
*Colocasia**	Elephant ear plant/Taro
*Conium maculatum**	Hemlock
*Convallaria majalis**	Lily of the Valley
Cordyline fruticosa	Hawaiian ti
Cordyline terminalis	Baby doll ti
*Coronilla varia**	Crown Vetch
Crocus	
Cycas revoluta	Sago palm
Cyclamen Persicum	Cyclamen
Daphne	
*Datura stramonium**	Thornapple/Jimsom weed

Sesbania punicea	Rattlebox, Purple Sesbane
Delairea odorata German Ivy	
*Delphinium**	Larkspur
*Dieffenbachia**	Dumb cane
*Digitalis purpurea**	Foxglove
Dracaena draco	Dragon tree
Dracaena fragrans	Corn plant/Dragon tree/Ribbon plant
*Epipremnum Aureum**	Devil's Ivy/Taro vine/Pothos
Equisetum	Horsetail
Eucalyptus	
Euonymus	Spindle tree
Euphorbia ingens	Cadelabra tree
Euphorbia marginata	Snow-On-The-Mountain
Euphorbia milii	Crown of Thorns
*Euphorbia pulcherrima**	Poinsettia
Euphorbia	Spurge
Euphorbia tirucalli	Pencil Cactus/pencil tree
Ficus benjamina	Weeping Fig
Ficus carica	Fig (sap)
Ficus elastica	Indian Rubber plant
Ficus lyrata	Fiddle leaf fig
Ficus microcarpa	Indian laurel tree
Galanthus	Snowdrop
Gelsemium sempervirens	Yellow jessamine
Geranium	
Ginkgo biloba	Ginko
Gladiolus	Gladiola
Gleditsia triacanthos	Honey Locust
Gloriosa	Glory Lilly
Gymnocladus dioicus	Kentucky Coffeetree
Gypsophila paniculata	Babies Breath
*Hedera helix**	Branching ivy/English ivy
Heliotropium	Heliotrope
Helleborus niger	Christmas Rose
Hemerocallis	Daylily
Hibiscus syriacus	Hibiscus
Hippophae rhamnoides	Buckthorn
Hura crepitans	Sandbox tree

*Hyacinthus**	Hyacinth
*Hydrangea**	
Hyoscyamus niger	Henbane
Ilex aquifolium	Holly
Impatiens	Jewelweed
*Ipomoea**	Morning Glory
*Iris**	
Jacobaea vulgaris	Ragwort/Tansy ragwort
Jatropha multifida	Coral Plant
*Juniperus**	Juniper
*Kalanchoe**	
*Kalmia latifolia**	Calico bush/Mountain laurel
Laburnum anagyroides	Golden Chain
Lamprocapnos spectabilis	Bleeding heart/Dutchmans breeches
Lantana camara	Lantana
Lathyrus odoratus	Sweet Pea
Leucojum aestivum	Snow Flake
*Ligustrum**	Privet
Lilium	Lily
*Lobelia**	
*Lobelia cardinalis**	Cardinal Flower
*Lolium perenne**	Perennial rye grass
Lonicera	Honeysuckle
Lophophora williamsii	Peyote
Lupinus	Lupine or Blue Bonnets
Lycopersicon esculentum	Tomato plant
*Macrozamia riedlei**	Zamia palm
Malus domestica	Apple (seeds)
Mandragora officinarum	Mandrake
Melia azedarach	Chinaberry tree/White cedar
Menispermum canadense	Moonseed
Mentha pulegium	European Pennyroyal
Mirabilis jalapa	Four O'clock
Momordica charantia	Balsam Pear
Monstera deliciosa	Ceriman/Fruit salad plant
Myristica	Nutmeg
Nandina domestica	Heavenly Bamboo
*Narcissus**	Daffodil

*Nerium oleander**	Oleander
*Nicotiana glauca**	Tobacco tree
*Nicotiana tabacum**	Tobacco
*Ornithogalum umbellatum**	Star of Bethlehem/Snowdrop
Oxalis	Shamrock plant
*Oxytropis**	Locoweed
Pachyrhizus erosus	Yam bean
Paeonia	Peony
Papaver	Poppies
*Parthenocissus quinquefolia**	Virginia Creeper
Parthenocissus tricuspidata	Boston ivy
Pastinaca sativa	Parsnip plant
*Persea americana**	Avocado
*Petroselinum**	Parsley
Philadelphus coronarius	Mock orange
*Philodendron**	
Philodendron Cordatum	Cordatum
Phoradendron villosum	Mistletoe
Physalis alkekengi	Chinese Lantern
*Phytolacca americana**	Pokeweed
Pieris japonica	
Podocarpus macrophyllus	Buddhist pine
Podophyllum peltatum	May apple
*Poinciana gilliesii**	Bird of Paradise
Polygonatum	Solomon's seal
Poncirus trifoliatia	Poncirus
Primula	Cowslip
Prunus armeniaca	Apricot (plant)
*Prunus caroliniana**	Carolina cherry laurel
*Prunus**	Cherry trees (plant, seed)
Prunus domestica	Plum (plant, seed)
Prunus persica	Peach (pits, plant)
Pteridium aquilinum	Bracken fern
Pyracantha	Firethorn
Pyrus	Pear (plant)
*Quercus**	Oak
Radermachera sinica	China doll/Serpent tree/Emerald tree
*Ranunculus**	Buttercup

*Rheum rhabarbarum**	Rhubarb
*Rhododendron**	
*Rhododendron occidentale**	Western azalea
Rhus radicans	Poison ivy/Poison sumac
*Ricinus communis**	Castor bean/Castor oil plant
*Robinia pseudoacacia**	Black Locust
Rumex acetosa	Sorrel
Salvia officinalis	Sage
Sambucus nigra	Black Elderberry (plant)
Sanguinaria canadensis	Bloodroot
*Sansevieria trifasciata**	Mother-in-law's tongue
*Schefflera**	
Schlumbergera truncata	Christmas Cactus
*Scindapsus Epipremnum**	Marble Queen
*Securigera varia**	Crown Vetch
Senecio rowleyanus	String of Pearls/beads
Senna alata	Christmas Candle
Senna obtusifolia	Java Bean/Glorybean
Sesbania punicea	Rattlebox/Purple Sesbane
*Sesbania vesicaria**	Bagpod/Bladder pod
Solandra maxima	Chalice vine
Solanum carolinense	Horse nettle
Solanum lycopersicum	Tomato (plant)
Solanum melongena	Eggplant (plant)
*Solanum**	Nightshades
*Solanum pseudocapsicum**	Jeruzalem cherry/Christmas cherry
*Solanum tuberosum**	Potato (plant)
Sophora secundiflora	Mescal bean
*Spathiphyllum**	Peace Lily
Stenanthium	Featherbells, death camas
*Strelitzia Reginae**	Bird of Paradise
Symphoricarpos albus	Waxberry
*Symplocarpus foetidus**	Skunk Cabbage
Syngonium podophyllum	Arrowhead plant/Nephthytis
*Taxus**	Yew
Toxicoscordion	Death camas
Toxidendron radicans	Poison ivy/Poison sumac
Trachelospermum asiaticum	Asiatic jasmine

Trachelospermum jasminoides	Confederate/Southern jasmine
*Trichodesma incanum**	
Tulipa	Tulip
Uncaria tomentosa	Catclaw
Veratrum viride	False hellebore
Verbena	Vervain
Vicia faba subsp. equina	Horse beans
Vicia	Vetch
Vinca	Periwinkle
Viscum album	Mistletoe
*Wisteria**	
Yucca	
*Zantedeschia**	Calla lily/Arum lily
Zigadenus	Death camas

*Are suspected of serious toxicity.

A11 – Psittacosis

Psittacosis is an infectious disease caused by *Chlamydia psittaci*. Psittacosis is spread from bird to bird, but it is also a zoonotic disease, meaning that the disease is transmissible from birds to humans. Symptoms in birds and humans can vary from mild to severe (including death).

Symptoms in birds can include dyspnea, diarrhea, discharge from eye or nose, conjunctivitis, neurological abnormalities, weight loss, ruffled feathers, green or yellow urates, and death. In most cases, symptoms of psittacosis appear within a few weeks after infection, but incubation time can take up to 2 years after the initial latent infection.

Practicing good hygiene is advised when handling and treating birds with signs of psittacosis. Samples of exudate, choana and cloaca taken with dry swabs can be tested for *Chlamydia psittaci* by polymerase chain reaction (PCR) testing.

In case of suspicion of psittacosis, start treatment with doxycycline immediately and keep the birds in quarantine pending results of the PCR test.

A12 – Disorders of calcium metabolism

Unbalanced diets leading to inadequate intake of calcium (low), phosphorus (high), vitamin D3 (low) and lack of exposure to sufficient UV-B light are common causes of disorders of calcium metabolism in pet birds. Excessive egg laying and kidney failure also have a negative impact on calcium metabolism.

Disorders of calcium metabolism can lead to hypocalcemia, egg binding (see Egg binding, p. 105), deformed bones in juvenile birds (osteodystrophy) and osteoporosis in adult birds, easily leading to fractures after minimal trauma (see Abnormal position of limbs: Fractures and luxations, p. 131)

Acute symptoms of hypocalcemia are caused by a decreased ionized blood calcium concentration). African grey parrots are especially sensitive for hypocalcemia; birds of other species seem to have better regulation of blood calcium concentration. Symptoms of hypocalcemia include general weakness, twitches, seizures, dyspnea and lack of muscle control or coordination.

X-RAYS

Thin skeletal cortices, (pathological) fractures, deformed bones, absence of polyostotic hyperostosis (increased bone density in the medullary cavities of the bones) in reproductively active hens and thin-shelled eggs can be noticed on X-rays in cases of disorders of calcium metabolism.

Note: Hypocalcemia can be an acute situation and absence of radiological abnormalities does not rule out hypocalcemia. On the other hand, radiological abnormalities consistent with problems with calcium metabolism do not prove hypocalcemia either, although they can be indicative.

BLOOD TESTS

Total blood calcium consists of calcium bound to proteins, calcium bound to minerals and ionized calcium. Only the free and active ionized calcium fraction is involved in calcium homeostasis. Hypocalcemia can only be diagnosed by measuring the level of free and active ionized calcium in the blood. Decreased ionized calcium indicates clinically relevant hypocalcemia. Increased ionized calcium indicates clinically relevant hypercalcemia. Abnormal total calcium is, in most cases, not caused by problems with calcium homeostasis. See Observation, physical examination and diagnostic tests.

MANAGEMENT

For the short term, calcium (boro)gluconate and vitamin D supplements can be administered. In birds with symptoms of hypocalcemia (e.g., weakness, ataxia, seizures, paresis, tremors, 'hiccups,' dyspnea), the initial administration of calcium (boro)gluconate is intramuscular Also, when blood has been taken for the determination of the (ionized) calcium concentration, treatment should be initiated immediately while awaiting results, as hypocalcemia can lead to rapid deterioration and death.

For the long term, diet and living condition should be optimized. The diet must contain sufficient calcium and vitamin D3 and have a favorable calcium/phosphorus ratio. In psittacines, feeding 75% of pelleted diets is a practical way to ensure sufficient calcium uptake. In addition to changes in diet, birds should be exposed to sufficient UV-B light (sunlight or artificial source/lamp). In cases of excessive egg laying, changes in husbandry or interaction with owners can be beneficial. Hormonal therapy can also be used to reduce reproductive activity.

A13 – First-aid kit at home

- Self-adhesive elastic bandage.
- Adhesive tape (for example, Leukoplast).
- Mosquito forceps.
- Povidone iodine.
- Physiological saline/NaCl.
- Silver nitrate stick.
- Styptic powder.
- Sterile gauze.
- Syringes.
- Handfeeding or recovery formula.
- Gloves.
- (Paper) towel.
- Peanut butter/Olive oil/Salad oil/Sunflower oil.
- Dextrose.
- Oral Rehydration Solution (ORS).
- Travel cage.
- Contact details from a nearby emergency clinic, avian specialist and poisoning information center.

A14 – Extra 'avian' veterinary materials

- Silver nitrate stick.
- Neck collars for birds in various sizes.
- Handfeeding or recovery formula for herbivores/liquid meat-based food for meat eaters.
- Crop tubes (metal and flexible) in various sizes.
- Stepped wall silicone endotracheal tubes 1–5 mm.
- Isoflurane/sevoflurane anesthetic system with see-through anesthetic masks in various sizes.
- Towels.
- Tissue glue.
- Baking soda.
- Oral Rehydration Solution (ORS).
- Adhesive plaster/tape (for example, Leukoplast).
- Epoxy resin.

A15 – Formulary

This formulary is not a complete list of drugs for birds. It is a list of the drugs mentioned in this book as used by the author. Please see James W. Carpenter's *Exotic Animal Formulary* (5th edition) for extensive information on medication for birds.

Antibiotic agents

- Amoxicillin/clavulanic acid 125 mg/kg PO q12h
- Azithromycin 40 mg/kg PO q24h
- Doxycycline 25 mg/kg PO q12h or 50 mg/kg IM q5–7d (IM causes muscle damage and pain)
- Enrofloxacin 15–30 mg/kg PO, IM q12h (IM causes muscle damage and pain)
- Metronidazole 50 mg/kg PO q24h or 25 mg/kg q12h
- Trimethoprim/sulfadiazine 50mg/kg PO q12h.

Antifungal agents

- Amphotericin B 100 mg/kg PO q12h 30 days (megabacteriosis/*Macrorhabdus ornithogaster*) or 1 mg/kg IT q12h (syringeal/tracheal aspergillosis) or 1 mg/ml nebulization during 15 minutes q12h (aspergillosis)
- Clotrimazole 10 mg/ml nebulization during 30–45 minutes q24h, 3 days on, 2 days off, repeat (aspergillosis); 2mg/kg q24h 5 days IT (syringeal/tracheal aspergillosis)
- Itraconazole 10 mg/kg PO q24h (toxicity reported in African grey parrots)
- Nystatin 300,000 IU/kg PO q12h 7–14 days
- Terbinafine 15 mg/kg PO q12h
- Voriconazole 15 mg/kg PO q12h.

Anthelmintic agents

- Fenbendazole 10 mg/kg PO q24h 3 days or 25–50 mg/kg PO once
- Flubendazole 1.43 mg/kg PO q24h 7 days
- Ivermectin 0.2 mg/kg PO, SC, IM
- Praziquantel 10 mg/kg PO.

Anticoccidial agents

- Sulfadimethoxine 20–50 mg/kg PO q12h 3–5 days
- Toltrazuril 7–15 mg/kg PO q24h 3 days.

Agents against flagellated protozoans

- Metronidazole 50 mg/kg PO q24h or 25 mg/kg PO q12h 3–10 days
- Ronidazole 6–10 mg/kg PO q24h 7 days.

Analgesic agents

- Butorphanol 1–4 mg/kg IM q2–4h
- Gabapentin 10–30mg/kg q8–12h
- Lidocaine (without adrenaline) 1–3 mg/kg
- Meloxicam 0.5–1.5 mg/kg PO/IM
- Tramadol 10–30 mg/kg q6–12h (use lower end of dosage range for birds of prey).

Antiemetic agents

- Maropitant 1 mg/kg SC, IM q24h
- Metoclopramide 0.3–0.5 mg/kg IM, IV, PO q8–12h.

Antiepileptic agents

- Levetiracetam 50 mg/kg PO q8h
- Midazolam 0.5–1.5 mg/kg IV, IM, IN
- Phenobarbital 1–7 mg/kg PO q12h.

Agents for cardiopulmonary resuscitation (CPR)

- Adrenaline/epinephrine 0.5–1 mg/kg IM, IV, IO, IT (cardiac arrest)
- Atropine 0.5 mg/kg IM, IV, IO (bradycardia)
- Doxapram 5–20 mg/kg IM, IV, IO (apnea).

Miscellaneous agents

- Activated carbon granulate 2,000–8,000 mg/kg PO
- Calcium (boro)gluconate 50–100 mg/kg IM/PO
- CaNaEdetate/CaNaEDTA 30–35 mg/kg IM q12h 3–5 days
- Dinoprostone gel 1 ml/kg cloacal
- Honey ointment topically q12h
- Lactulose 150–650 mg/kg PO q8–12h
- Midazolam 0.5–1.5 mg/kg IV, IM, IN
- Penicillamine 50–55 mg/kg PO q24h 3–6 weeks
- Salbutamol 0.1–0.2 mg/kg IM or nebulization
- Sucralfate suspension 25 mg/kg PO q8h
- Vitamin D3 3,300 IU/kg IM
- Silver sulfadiazine topically q12–24h.

A16 – Biochemistry reference intervals

The following tables list reference intervals for selected avian species. Data are adapted from *Exotic Animal Formulary* (5th edition, James W. Carpenter). Data with an asterisk (*) are adapted from the article *'Biochemical Reference Intervals for Backyard Hens'* (Melissa M. Board et al.), *Journal of Avian Medicine and Surgery* 32(4):301–306, 2018.

To save time while interpreting biochemistry results, both US Conventional Units and SI units are listed for some measuring units.

	African grey parrot	Cockatoos	Macaws	Caique	Amazon parrots
AST (U/L)	109–305	117–314	105–324	193–399	141–437
CK (U/L)	228–322	106–305	101–300	134–427	125–345
Bile acids (umol/L)	12.0–96	34–112	7.0–100	12.0–112	33.0–154
Uric acid (mg/dL)	2.7–8.8	2.9–11.0	2.9–10.6	3.4–12.2	2.1–8.7
Uric acid (umol/L)	161–523	172–654	172–630	202–726	125–517
BUN (mg/dL)	3.0–5.4	3.0–5.1	3.0–5.6		
BUN (mmol/L)	1.1–1.9	1.1–1.8	1.1–2.0		
Total protein (g/L)	32–52	30–50	26–50	24–46	30–52
Albumin (g/L)	12.2–25.2	11.1–22.8	11.2–24.3	9.6–20.4	17.9–28.1
Calcium (mg/dL)	7.7–11.3	8.3–10.8	8.2–10.9	7.1–11.5	8.2–10.9
Calcium (mmol/L)	1.9–2.8	2.1–2.7	2.0–2.7	1.8–2.9	2.0–2.7
Glucose (mg/dL)	206–275	214–302	228–325	167–366	221–302
Glucose (mmol/L)	11.4–15.3	11.9–16.8	12.7–18.0	9.3–20.3	12.3–16.8
Potassium (mEq/L)	2.9–4.6	2.5–4.5	2.0–5.0		3.0–4.5
PCV (%)	45–53	40–54	42–56	47–55	41–53
WBC (x10^3/uL)	6.0–13.0	5.0–13.0	10.0–20.0	8.0–15.0	6.0–17.0

	Cockatiel	*Agapornis* spp.	Budgerigar	Eclectus parrot	Pigeon
AST (U/L)	160–383	125–377	55–154	148–378	45–123
CK (U/L)	58–245	58–337	54–252	118–345	110–480
Bile acids (umol/L)	44–108	12.0–90	32–117	30–110	22–60
Uric acid (mg/dL)	3.5–11	2.5–12	3.0–8.6	2.5–8.7	2.5–12.9
Uric acid (umol/L)	208–654	149–714	178–511	149–517	149–767
BUN (mg/dL)	2.9–5	2.8–5.5	3.0–5.2	3.5–5.0	2.4–4.2
BUN (mmol/L)	1.0–1.8	1.0–2.0	1.1–1.9	1.2-–1.8	0.9–1.5
Total protein (g/L)	24–48	24–36	20–30	30–50	21–33
Albumin (g/L)	7.8	9.8–16.8	17.5	12.3–22.6	13–22
Calcium (mg/dL)	7.3–10.7	7.2–10.6	6.4–11.2	7.9–11.4	7.6–10.4
Calcium (mmol/L)	1.8–2.7	1.8–2.6	1.6–2.8	2.0–2.8	1.9–2.6
Glucose (mg/dL)	249–363	246–381	254–399	220–294	232–369
Glucose (mmol/L)	13.8–20.2	13.7–21.1	14.1–22.1	12.2–16.3	12.9–20.5
Potassium (mEq/L)	2.4–4.6	2.1–4.8	2.2–3.7	3.5–4.3	3.9–4.7
PCV (%)	43–57	44–55	44–58	45–55	49±3.8
WBC (x103/uL)	5.0–11.0	7.0–16.0	3.0–10.0	9.0–15.0	

	Chicken	Harris's hawk	Peregrine falcon
AST (U/L)	118–298*	95–210	20–52
CK (U/L)	107–1,780*	224–650	357–850
Bile acids (umol/L)	<45*		20–118
Uric acid (mg/dL)	2.5–8.1	9–13.2	4.4–22
Uric acid (umol/L)	149–482	535–785	262–1,309
BUN (mg/dL)			
BUN (mmol/L)			
Total protein (g/L)	33–55	31–46	25–40
Albumin (g/L)	13–28	14–17	8.0–13.0
Calcium (mg/dL)	13.2–23.7	8.4–10.6	8.4–10.2
Calcium (mmol/L)	3.3–5.9	2.1–2.6	2.1–2.5
Glucose (mg/dL)	227–300	220–283	
Glucose (mmol/L)	12.6–16.7	12.2–15.7	
Potassium (mEq/L)	3.0–7.3	0.8–2.3	1.6–3.2
PCV (%)	23–55	32–44	37–53
WBC (x103/uL)	9.0–32.0	4.8–10.0	3.3–21

A17 – Anatomy

Figs. A17.1-A17.5 show schematic drawings of the digestive tract, respiratory tract, female and male urogenital tract and skeleton of an African grey parrot.

Note: Anatomy varies between species. For example, many birds have caeca and some birds (including owls and waterfowl) don't have a real crop.

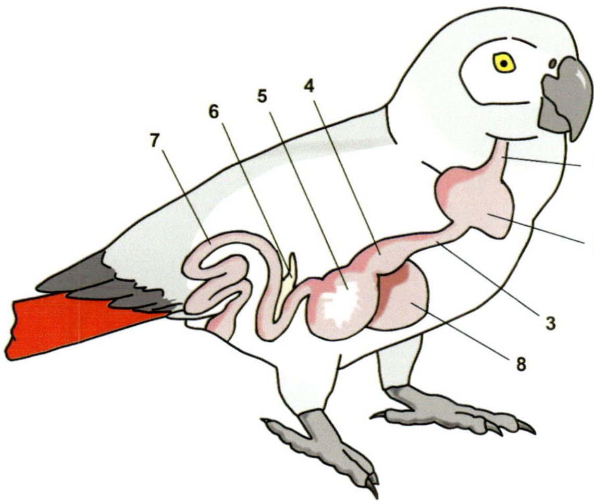

Fig. A17.1 Schematic drawing of the digestive tract: (1) Cervical oesophagus, (2) crop, (3) thoracic oesophagus, (4) proventriculus, (5) ventriculus, (6) pancreas, (7) intestine and (8) liver.

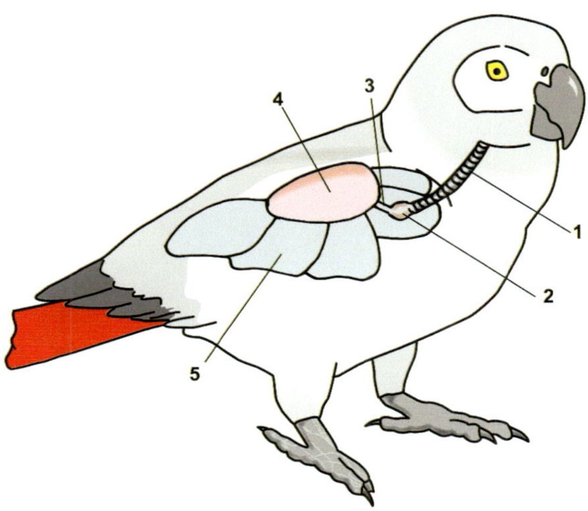

Fig. A17.2 Schematic drawing of the lower respiratory tract: (1) Trachea, (2) syrinx, (3) bronchus, (4) lung and (5) air sac.

Fig. A17.3 Schematic drawing of the female urogenital tract: (1) Ovarium, (2) oviduct, (3) uterus/shell gland, (4) intestine, (5) cloaca, (6) kidney and (7) ureter.

Fig. A17.4 Schematic drawing of the male urogenital tract: (1) Testis, (2) kidney, (3) vas deferens and ureter and (4) cloaca.

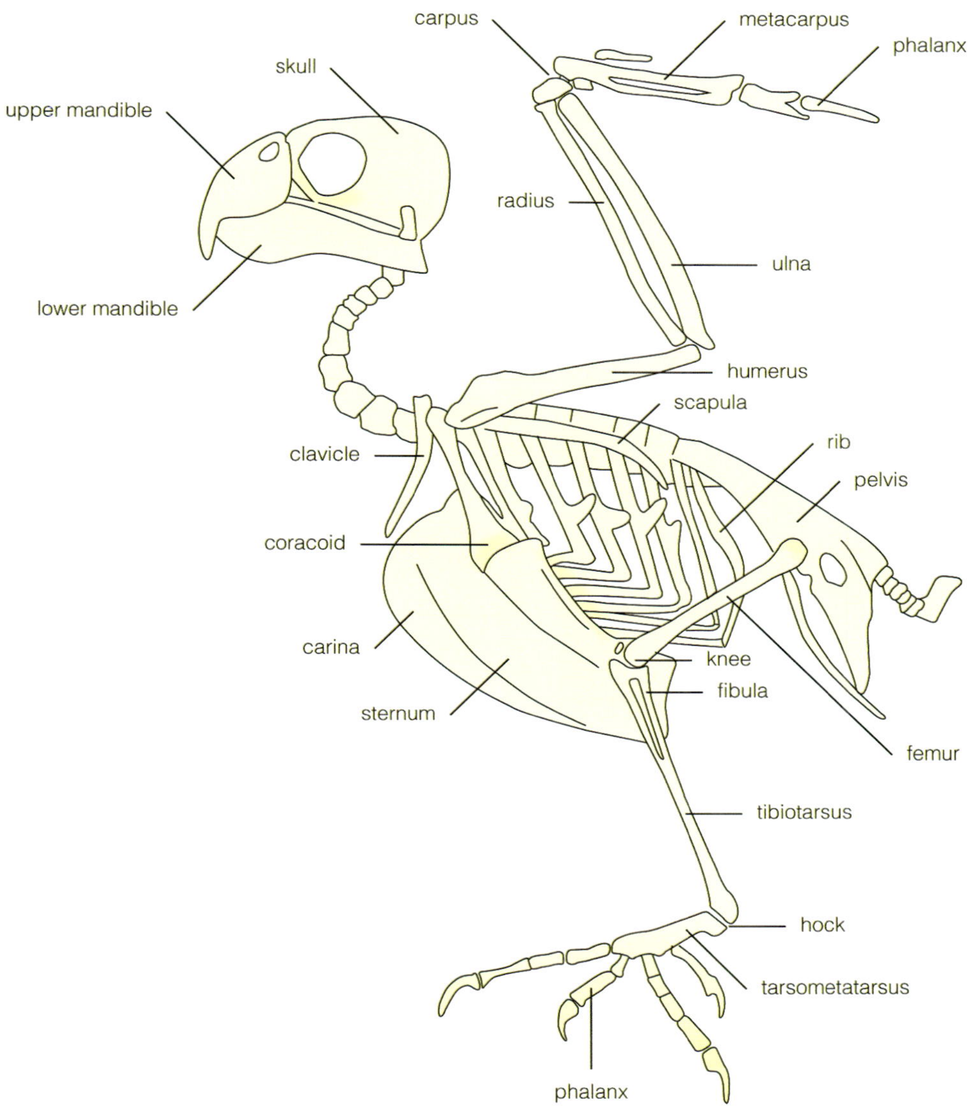

Fig. A17.5 Schematic drawing of the skeleton.

Index

For Product Safety Concerns and Information please contact our EU representative GPSR@taylorandfrancis.com Taylor & Francis Verlag GmbH, Kaufingerstraße 24, 80331 München, Germany

Batch number: 10733635

Printed by Printforce, the Netherlands